COMMUNITYBUSHFIRESAFETY

EDITORS: JOHN HANDMER AND KATHARINE HAYNES

CSIRO
PUBLISHING

National Library of Australia Cataloguing-in-Publication entry
Community bushfire safety.

Bibliography.
Includes index.
ISBN 9780643094260 (pbk.).

1. Fire management – Australia. 2. Forest fires –
Prevention and control – Australia. 3. Dwellings – Fires
and fire prevention – Australia. 4. Wildfires – Prevention
and control. 5. Sociology, Rural – Fires and fire
prevention – Australia. I. Handmer, J. W. (John William),
1950– . II. Haynes, Katharine, 1977– . III. CSIRO
Publishing.

363.3770994

Published by
CSIRO PUBLISHING
150 Oxford Street (PO Box 1139)
Collingwood VIC 3066
Australia

Telephone: +61 3 9662 7666
Local call: 1300 788 000 (Australia only)
Fax: +61 3 9662 7555
Email: publishing.sales@csiro.au
Website: www.publish.csiro.au

Cover
Bush paintings by Mark Schaller, an Australian artist based in Melbourne, who exhibits regularly. These works are the result of the devastating fires in the Grampians. (mschaller@bigpond.com)

Front cover: 'Smoldering Hill' 2006 oil on linen, 186 × 137 cm.
Back cover: 'After the Fire' 2006 oil on linen, 186 × 137 cm (left); 'Mount William (Grampians)' 2006 oil on linen, 186 × 137 cm (centre); 'Re-growth Silverband Falls' 2006 oil on linen, 186 × 137 cm (right).

Set in 10pt/12pt Minion
Edited by Adrienne de Kretser
Cover and text design by James Kelly
Typeset by Palmer Higgs
Printed in Australia by Ligare

Contents

Foreword: a view from Australia   v
John Gledhill

Foreword: a view from North America   vii
Jerry Williams

Statement: Bushfire Co-operative Research Centre   ix
Kevin O'Loughlin

Acknowledgments   x

Author biographies   xi

Interface bushfire community safety

Chapter 1: Interface (urban–rural fringe) bushfire community safety   3
John Handmer and Katharine Haynes

Understanding communities

Chapter 2: Community perceptions of bushfire risk   11
Alison Cottrell, Sally Bushnell, Margaret Spillman, Judy Newton, David Lowe and Luke Balcombe

Chapter 3: Resilience at the urban interface: the Community Fire Unit approach   21
Tom Lowe, Katharine Haynes and Gerry Byrne

Chapter 4: The concept of local knowledge in rural Australian fire management   35
Jenny Indian

Chapter 5: Social contexts of responses to bushfire threat: a case study of the Wangary fire   47
Helen Goodman and Mae Proudley

Assisting the householder and small business operator

Chapter 6: Prepare, stay and defend or leave early: evidence for the Australian approach   59
Amalie Tibitts, John Handmer, Katharine Haynes, Tom Lowe and Josh Whittaker

Chapter 7: Property safety: judging structural safety   77
Raphaele Blanchi and Justin Leonard

Chapter 8: Don't get burnt by the law: the legal implications of the 'prepare, stay and defend or leave early' policy   87
Elsie Loh

Risk prevention and communication

Chapter 9: Understanding and preventing bushfire arson   99
Damon Muller and Colleen Bryant

Chapter 10: The media and fire services: dealing with conflicting agendas   107
Erez Cohen, Peter Hughes and Peter B. White

Chapter 11: Preparing for bushfires: the public education challenges facing fire agencies   117
Douglas Paton and Lyndsey Wright

Policy and institutional issues

Chapter 12: Using program theory in evaluating bushfire community safety programs   129
Alan Rhodes and John Gilbert

Chapter 13: What should community safety initiatives for bushfire achieve?   139
Gerald Elsworth, Karl Anthony-Harvey-Beavis and Alan Rhodes

Chapter 14: The economics of bushfire management   151
Gaminda Ganewatta

Chapter 15: Save that brigade! Recruiting and retaining fire service volunteers to protect your community   161
Jim McLennan, Adrian Birch, Sean Cowlishaw and Joel Suss

The future with a warmer climate

Chapter 16: Climate change and community bushfire resilience   175
Karyn Bosomworth and John Handmer

References   185

Index   201

Foreword
A view from Australia

Over the last 50 years Australian emergency service agencies have steadily developed in capability and sophistication. In consequence, there seems to be increasing community reliance on them at times of emergency or crisis and a corresponding decrease in self-reliance, particularly in urban and semi-urban communities.

Australian communities, especially in rural areas, were historically very self-reliant by necessity. Although there were some rudimentary rural fire brigades, home and asset protection from bushfires were the responsibility of individuals. People had a fundamental responsibility for their own safety. Bushfires were part of life and people did their best to protect themselves and their properties from fires. The science of fire behaviour was not well advanced and, in the absence of effective firefighting technology, there were frequently severe losses of both property and lives from bushfires.

Research has informed our knowledge of fire behaviour, the mechanisms of ignition of buildings, and how people were killed and injured by bushfires. Although bushfires are often portrayed in the media as irrational demons, we know that they behave and respond in predictable ways to the environmental variables of fuels, weather and slope. This means we should be more effective in managing the safety of people and their property from bushfires. This increased knowledge, coupled with improved firefghting organisations, firefighting technology and communications, have made bushfire management more effective and successful. But still Australia suffers from increasingly frequent bushfires, ranging from damaging to disastrous.

In recent years Australian fire agencies have realised that bushfire management challenges are growing as urban environments expand into the bush. The number of people and properties in areas at-risk from bushfire is steadily expanding. Research carried out before the inception of the Bushfire Co-operative Research Centre clearly established that:

- most people who perished in bushfires died in the open or escaping in cars or on foot
- most houses lost in bushfires were unoccupied, i.e. no one was at home to protect them
- windborne embers are the main ignition source in buildings damaged or lost during bushfires
- householder preparations (particularly creating a fuel-reduced area adjacent to buildings) are critical in house survival
- bushfires under severe conditions can simultaneously affect many properties, overwhelming firefighting resources.

These factors led fire agencies to promote the efficacy of householder preparation and to encourage residents to stay and defend their properties. This strategy has wide acceptance and national support in an Australasian Fire Authorities Council position paper.

It requires a fundamental shift in responsibility for safety from the firefighting agency to those at risk. This is a significant shift that involves many questions. It requires an understanding of many aspects of different communities and human behaviour so that fire agencies can assist and empower people to manage their bushfire safety.

The human factors in relation to bushfires have not had the same research attention. With the establishment of the Bushfire Co-operative Research Centre, a new program of research began to investigate many facets of the human/community response to bushfire. Until its establishment fire agencies began to implement the new bushfire strategy without the benefit of much significant relevant research. The Bushfire CRC's Research Program *Community Self-sufficiency for Fire Safety* is well underway and is providing assistance to fire agencies.

A crucial activity of the Bushfire CRC involves transferring research findings to fire agencies, to influence practical implementation. Fire agencies are eager to access research results that inform them about a range of issues relating to the empowerment of at-risk communities. This book is a very effective way of transferring such knowledge.

Social science research is a relatively new area for fire agencies in Australia, and its value is obvious. Fire agencies protect life, property and the environment. Fire research in the past has mainly been focused on fire and firefighting, but fire agencies' protection priority is the community/people. It therefore seems logical that we need to gain a better understanding of communities, people and their needs and perceptions, if we are to engage with them so that they can become more self-reliant with respect to their safety from bushfire.

It is generally accepted that in response to climate change our environment is likely to suffer more frequent bushfires, some of great intensity. I believe that improved firefighting technology alone cannot provide a way of dealing with the new fire regime. The change of paradigm, where communities at risk accept greater responsibility for their own safety, offers a much more successful, sustainable and affordable solution. Creating this self-reliance requires a greater understanding of the issue.

Bushfire CRC researchers have a very obvious passion and realise that they are undertaking something that is really making a difference. I commend their efforts and this book as a great start in a journey of supporting a new paradigm in bushfire risk management.

John Gledhill
Chief Officer
Tasmania Fire Service

Foreword
A view from North America

Every year, bushfires burn across large portions of Australia, New Zealand, the Mediterranean countries and the US. In these parts of the world, bushfire seasons have become longer and more severe. Despite rapid and aggressive firefighting actions, some incidents become mega-fires. They exceed all efforts at direct control until there is a break in fuel or change of weather. These rare but high-consequence bushfires often become 'record-breakers'. The suppression costs, property losses and environmental damage, including to watersheds and biodiversity, have become alarming. International fire services are increasingly recognising that climate change, over-accumulated fuels in forested ecosystems, and unprecedented growth in human assets and activity near the wildland–urban interface underlie the mega-fire phenomenon.

Fire professionals are also coming to realise that simply answering an increasing bushfire threat with greater suppression force – at the tactical level – is no longer enough. In some places, they are beginning to complement firefighting strength with a broader focus on strategic solutions that aim at improving the health and resilience of fire-prone bushlands and positively influencing how people think about living in the wildland–urban interface.

In 1998, during a professional exchange between Australia, New Zealand and the US, a group of bushfire experts began discussing the social science aspects of mega-fires. It was pointed out that, although much progress has been made in fire behavior modelling, fuel inventory techniques, fire weather forecasting, apparatus engineering and other facets of the physical sciences, we knew little about the people and community side of bushfire management. In recent years, much progress has been made through community-based bushfire protection programs, including Firewise planning in the US. In some places, community leaders, policy-makers and law-makers have worked together to significantly improve the safety of citizens, the protection of property and the sustainability of the forest. But given the enormity of the bushfire threat, progress has been slow and successes have been limited.

In 2002, Australia initiated the first Co-operative Research Centre (CRC) focused on the bushfire problem. Major fires in New South Wales (2001/02), Canberra and the Alps (2003) and the Eyre Peninsula and Perth Hills (2005) underscore the growing bushfire threat. The social dimension of dealing with the bushfire threat is a core element of the CRC.

I was privileged to be a part of a panel that conducted a mid-course review of the Bushfire CRC. I was also privileged to take part in the National Bushfire Forum, held in Canberra during February 2007. The work that is being accomplished under Program C (Community Self-sufficiency) of the Bushfire CRC (reported in this book), is a most positive contribution to the social dimension of bushfires. The work that is reflected here is among the most progressive and innovative anywhere in the world.

In many respects, the mega-fire phenomenon is a new reality. Our countries have a history of large destructive bushfires, but the convergence of global warming, fuel build-ups and expanding growth in and around fire-prone forests suggests that mega-fire consequences will increase unless fire-dependent ecosystems are better managed, people are more aware and communities are better prepared. Australia's investment in bushfire research through the Co-operative Research Centre signals the country's commitment to becoming a world leader in dealing with the bushfire threat and the reality of mega-fires.

Jerry Williams
Senior Policy Advisor, Megafire Project, Brookings Institute
Former National Fire Director
US Forest Service

Statement
Bushfire Co-operative Research Centre

Research on the community safety aspects of bushfires provides fascinating insights to key bushfire and land management policies and practices in the future. It was an exciting and innovative part of the research agenda for the Bushfire Co-operative Research Centre's program when it got underway in mid 2003.

The two main Bushfire CRC programs concerned with people are 'Community Self-sufficiency for fire safety', led by Professor John Handmer, and 'Protection of people and property', led by Dr Bob Leicester. Even before the research commenced, Australia and the world were given a stark reminder of the potential of severe bushfires to wreak havoc on communities. On 18 January 2003 a bushfire swept into the suburbs of Canberra, causing the loss of four lives and destruction of over 500 houses and other buildings.

This event encapsulated many of the people issues and paradoxes of bushfires at the urban interface. What did the community understand about the threat on that day? How many of the houses were unoccupied due to the residents being away on summer holidays? How did one of the CRC's senior research fellows, Dr Malcolm Gill, single-handedly defend his house that directly faced the fire in ferocious winds? Why did some houses that were one or two streets back from the interface burn, when those directly at the interface survived?

Later that same year in the northern hemisphere summer, over 20 people died in extreme fires in Portugal, there were severe fires in France and Spain and, in October 2003, fires in southern California destroyed over 3000 homes. The record fires in both hemispheres in one year raised questions about links with climate change. Were communities in fire-prone countries going to face more frequent severe fires in the future? How much would this be exacerbated by the rapid expansion of cities into the bush?

The concentration of these major fires in time, but in diverse geographic locations, focused increased international attention on the Australasian research, especially the issues of community, people and property. The reason for this interest was the novel approach taken by Australian fire authorities to promote community resilience and reduce bushfire risk. For example, the concept of able-bodied, well-prepared people staying in their homes and defending them in the direct face of a fire front was almost unheard-of in the US, but is now being examined there because of the Australian fire agencies' policies and recent research supporting this approach.

The chapters in this volume form a rich sample of the work being conducted by a range of research partners in the Bushfire CRC. The research is guided by the needs of the Australasian fire and land management authorities. It is funded by the contributions of cash and in-kind resources from our partners and an annual cash grant from the Australian government through the CRC program that spans the funding term of 2003–10.

This book presents the work of a team of dedicated, enthusiastic and passionate researchers who show great skill, initiative and sensitivity in their dealings with the community safety aspects of bushfires.

Kevin O'Loughlin
Chief Executive Officer
Bushfire Co-operative Research Centre

Acknowledgments

This book was developed to bring together the wide range of social science research undertaken within the Bushfire Co-operative Research Centre (CRC). To the best of our knowledge, it is the only such collection in the field of bushfire (or wildfire) research and practice. This book makes the material accessible to both non-specialists and bushfire researchers.

All the work reported here was undertaken as part of the Bushfire CRC program. We thank all those who sponsor and support Bushfire CRC research as funders, practitioners and colleagues, in particular Kevin O'Loughlin (CRC CEO), John Gledhill (End-user leader Program C) and Len Foster, who was a prime driver of the CRC idea.

We are grateful for the support and advice of our editorial committee. In addition to the book's editors, the Editorial Committee consisted of Damian Killalea and John Gledhill (Tasmania Fire Service); Naomi Brown (now CEO of the Australasian Fire Authorities Council, then with the Victorian Country Fire Authority and end-user leader); Richard Thornton (Research Director for the Bushfire CRC); and Kellie Watson (Bushfire CRC). Richard Thornton and Kellie Watson organised publication with CSIRO Publishing.

As an edited collection, the book depends on the efforts of its contributing authors and on the dedication of the reviewers. All chapters have been reviewed from a number of perspectives by the editors; two agency officials who work in research and community safety – Damian Killalea (Head Community Safety, Tasmania Fire Service) and David Nicholls (Research Co-ordinator for the CFA); and by Richard Thornton (of the CRC). Chapters that contained specific case studies or where the case was the subject of a public inquiry were also examined by fire and land-management agency staff familiar with the cases. Authors were invited to improve their chapters based on feedback from these reviewers and the quality of the material in this volume increased significantly as a result.

John Handmer
Katharine Haynes
Centre for Risk and Community Safety
RMIT University
Melbourne

Author biographies

Karl Anthony-Harvey-Beavis was research officer for the Bushfire CRC Evaluation Framework project in the Collaborative Institute for Research, Consulting and Learning in Evaluation (CIRCLE) at RMIT University, responsible for the data management and detailed analysis of the concept-mapping work for this project. Karl currently works for the Australian Crime Commission.

Luke Balcombe studied for his Environmental Science Masters degree at James Cook University and the Bushfire Co-operative Research Centre during 2004–06. The Tamborine Mountain Bushfire Awareness case study formed the basis of Luke's masters thesis, 'Perceptions of preparedness for bushfire'. Previously, Luke worked as a park ranger at Tamborine Mountain and as a research fellow at the World Forest Institute based in Portland, Oregon, where he investigated fire management on public lands. His areas of research interest include the environmental, economic, social and cultural aspects of fire management.

Adrian Birch is a research officer in the School of Psychological Science at La Trobe University, working on the Bushfire Co-operative Research Centre Enhancing Volunteerism Project. His background is in sociology. He is a long-serving CFA volunteer for the Country Fire Authority, Victoria.

Dr Raphaele Blanchi is a Bushfire CRC-funded post-doctoral research scientist at CSIRO Sustainable Ecosystems in Australia. She obtained a PhD in Science and Engineering of Hazardous Activities from Ecole des Mines de Paris, where she worked on forest fire risk assessment and prevention at the urban interface. Her interests lie in understanding the specific issues involved in house loss, and providing specific tools to predict and manage the risk posed by bushfires at the rural–urban interface.

Karyn Bosomworth spent several years in the community safety section of Victoria's Country Fire Authority, co-ordinating post-bushfire research assessing policy efficacy, among other projects. More recently with Victoria's Department of Sustainability and Environment, she has been working on climate change adaptation. She is currently on leave from DSE and undertaking her PhD on climate change and community resilience bushfire.

Sally Bushnell is a research officer for Bushfire CRC's Understanding Communities Project in the Centre for Disaster Studies at James Cook University. Sally completed her Master of Applied Science in Natural Resource Management at James Cook University in 2005. Her areas of research interest broadly include hazard and environmental management, in particular community participation in management and developing sustainable outcomes. Current research focuses on community and fire brigade perceptions, values and expectations relating to bushfire hazard.

Dr Colleen Bryant works with the Australian Institute of Criminology's bushfire arson project. She is principally involved with the analysis of spatial and temporal trends of deliberate firesetting in Australia.

Superintendent Gerry Byrne is the Manager of Bushfire Natural Hazards with the NSW Fire Brigades, a position he has held since 2005. Gerry has 24 years firefighting experience in metropolitan Sydney and regional NSW. He is responsible for the strategic management of the Community Fire Unit Program which has over 6000 volunteers, and has recently been appointed to the Fire Services Joint Standing Committee. He is finishing his Masters of Emergency Management, with a major focus on 'I-Zone' firefighting.

Dr Erez Cohen is a social anthropologist and has been a postdoctoral fellow in the media studies program at La Trobe University. He conducted Bushfire CRC-funded research on the relationship between emergency organisations and the media in Australia.

Dr Alison Cottrell is a senior lecturer in the Centre for Disaster Studies in the School of Earth and Environmental Sciences at James Cook University in Townsville, Queensland. Alison's research focus is communities and their relationships with hazards. Perceptions of risk, vulnerability and resilience at the individual, household and community levels are her prime concern. Current research includes a project with the Bushfire CRC on understanding the relationships between government policy, planning, service delivery and community responses to bushfires. This research is linked to an assessment of appropriate community-based strategies for enhancing community resilience to hazards in general. Alison's teaching activities include cultural aspects of environmental issues, disaster studies, community studies and qualitative social research methods.

Sean Cowlishaw obtained a Bachelor of Social Science with first-class honours in psychology from Swinburne University of Technology. He is a PhD candidate in the School of Psychological Science at La Trobe University, where he is researching the impacts of fire service volunteering on the families of volunteers.

Dr Gerald Elsworth is principal researcher in the Collaborative Institute for Research, Consulting and Learning in Evaluation (CIRCLE) at RMIT University and leader of the Bushfire CRC Evaluation Framework Project. He has worked in public sector policy research and program evaluation since the mid- 1970s, mainly in education and public health, but more recently in various aspects of community safety. His particular interests are the development of quantitative measures of program outcomes and the use of program theory in evaluation.

Dr Gaminda Ganewatta is an economist by training and currently serves as research fellow in Economics of Community Safety at the Centre for Risk and Community Safety, RMIT University. His research program examines the economic impact of bushfires on local, state and national economies and evaluates economic efficiency of various fire management strategies. He holds a BSc and MSc (Agricultural Economics) from the University of Peradeniya, Sri Lanka and a PhD from La Trobe University, Australia.

John Gilbert is a researcher with the Bushfire CRC, based at the Collaborative Institute for Research, Consulting and Learning in Evaluation (CIRCLE), RMIT University. His work has included several evaluations of specific bushfire community education, awareness and engagement programs, including Street FireWise in the Blue Mountains in NSW.

Dr Helen Goodman is a post-doctoral research fellow at RMIT University and brings over three decades of professional practice in social work, including research and evaluation work, to bear on understanding community responses to the threat of fire. As part of the Community Self-sufficiency research program of the Bushfire CRC, in particular C7, the Evaluation

Framework project, she is working to understand the interface between formal services and the community and contributing to the strengthening of dialogue at this interface.

Professor John Handmer is Innovation Professor in Risk and Sustainability at RMIT University in Melbourne. He is head of the Centre for Risk and Community Safety, directs RMIT's Human Security Program, and is a Research Program Leader for the Bushfire CRC. He is also an Adjunct Professor at the Australian National University. His research is dedicated to improving community resilience, safety and security through providing the evidence base for policy and practice.

Dr Katharine Haynes completed a PhD in risk management at the School of Environmental Science, University of East Anglia, UK, and is currently employed by Risk Frontiers, Macquarie University, Sydney. Katharine is also a trained wildland firefighter, having volunteered for the USDA Forest Service in Washington state. Her research interests include risk communication, disaster risk reduction and participatory processes. Katharine is working in collaboration with the Bushfire CRC and Centre for Risk and Community Safety on a number of projects relating to the 'Prepare, stay and defend or leave early' policy.

Dr Peter Hughes lectures in Media Studies at La Trobe University. His research and teaching focuses on documentary and new media. He is co-author, with Ina Bertrand, of *Media research methods: institutions, texts, audiences* (Palgrave, New York, 2005).

Jenny Indian has worked for thirty years as a landscape architect both within Australia and abroad and holds a Bach. App. Sc., Dip. Landscape Design and Grad. Dip. V.E.T. (Hons). Much of this work has involved consideration of the relationship of rural communities to landscape. For the past eighteen years Jenny has been based in north-east Victoria where she gained first-hand experience of the 2003 alpine fires and their impact on rural communities. She is currently researching the resilience of rural communities to bushfire, work undertaken as part of the HighFire Project.

Justin Leonard is a mechanical engineer leading the Bushfire Urban Design research area in CSIRO Sustainable Ecosystems. This work focuses on the delivery of risk assessment tools and urban design solutions for major clients and community education initiatives. He is working with the Bushfire CRC on a range of projects focusing on urban interface design.

Elsie Loh graduated in Law (Hons) and Science from Monash University and practised as a immigration lawyer before working as a legal research officer at the Centre for Risk and Community Safety. Her research areas include law in bushfire, vulnerability and law, refugee and immigration law, human rights law, law and development and comparative law. She volunteers as an Australian legal practitioner and a registered migration agent at the Refugee and Immigration Legal Centre. She is currently completing a Master of Laws at Melbourne University.

David Lowe is a research officer in the Bushfire CRC in the Centre for Disaster Studies at James Cook University. David has a BAdmin (Tourism) (Hons). His areas of research interest include social and political representations, employment discrimination, community perceptions and evaluation of vulnerability to hazards. Current research includes investigating the perceptions, values and expectations of communities and fire services.

Tom Lowe holds an honours degree in Forestry Management from the University of Aberdeen, Scotland, and a Masters in Environment and Development from the University of East Anglia. He has practical experience of forest operations and is a trained wildland firefighter, having spent two fire seasons as a volunteer for the USDA Forest Service in Washington state. For the past four years Tom has researched environmental decision-making, in particular public perceptions of environmental risks and the development of appropriate risk communication strategies.

Dr Jim McLennan is a senior research fellow in the School of Psychological Science at La Trobe University and manager of the Bushfire CRC Enhancing Volunteerism Project. His background is in organisational psychology.

Dr Damon Muller completed a doctorate in criminology at the University of Melbourne. He has worked at the Australian Institute of Criminology since 2006, initially as research fellow for the Criminology Research Council then as program leader for the bushfire arson project. His research interests and publications include juvenile justice, domestic violence, criminal profiling and homicide.

Judy Newton is Research Manager at Queensland Fire and Rescue Service in Brisbane, Queensland. Judy has managed a number of community safety research projects targeting at-risk and priority groups (indigenous youth, African refugees, people with disabilities, alcohol- and drug-dependent people) as well as employee attitude surveys, customer satisfaction surveys and behavioural research projects. Current research includes two research projects of national significance, namely the AFAC Accidental Fire Deaths and Injuries Project and the Bushfire CRC Understanding Communities Project with Dr Alison Cottrell from James Cook University, Townsville.

Dr Douglas Paton is a professor in the School of Psychology, University of Tasmania, a research fellow at the Institute of Geological and Nuclear Sciences, New Zealand, and research associate with Aragon Law, Perth, WA. His research focuses on developing and testing models of personal, family and community resilience for natural hazards and working in extreme environments. An all-hazards approach is adopted, and he is currently studying adaptive capacity in relation to bushfire, volcanic, seismic, tsunami and flu pandemic hazards. He was a member of a working group consulting to NATO to develop a European strategy for community resilience to terrorism and was Australian delegate to the UNESCO Education for Natural Disaster Preparedness in the Asia-Pacific program.

Mae Proudley obtained a BA with first-class honours from the University of Melbourne. She is a masters by research candidate at RMIT University, where she is undertaking a case study of the Wangary Fire for the Bushfire CRC.

Alan Rhodes is the Manager of Research and Evaluation at the Country Fire Authority in Victoria. He is currently undertaking a PhD at RMIT University on the evaluation of community safety programs for bushfire, as part of the Bushfire CRC.

Margaret Spillman is a research officer for Bushfire CRC's Understanding Communities Project in the Centre for Disaster Studies at James Cook University. Margaret has a BSc (Geography) (Hons), and is broadly interested in social studies in the area of hazard and risk management. Her current research investigates community perceptions and preparations for the bushfire hazard in peri-urban areas.

Joel Suss was awarded first-class honours in psychology at La Trobe University. He is a research assistant working on Bushfire CRC projects at La Trobe University.

Amalie Tibbits holds a BSc in Geography from the University of Melbourne. She has worked at the Centre for Risk and Community Safety on a variety of projects focusing on community participation in emergency management and the social impact of disasters. She lives in north-east Victoria and is part of the Highfire project of the Bushfire CRC, investigating aspects of community resilience to fire in the high country. Her research has also focused on documenting the evidence for the 'prepare, stay and defend or leave early' policy and identifying obstacles to implementation through a number of regional case studies.

Dr Peter B. White is conducting research on the social implications of new media and telecommunications policy. He is the Director of the La Trobe University Online Media Program which conducts research into the uses of new media. Peter directs the Media Studies program's television industry training program in Vietnam.

Josh Whittaker obtained a BSocSci (Environment) with first-class honours from RMIT University. He is a PhD candidate in the Centre for Risk and Community Safety at RMIT, where he is researching human vulnerability and resilience to bushfires in remote rural East Gippsland. Josh was awarded a full scholarship from the Bushfire CRC.

Lyndsey Wright was corporate planner at the Metropolitan Fire and Emergency Services Board (MFB) for ten years and recently took on the additional role of Manager Planning and Research, thereby linking research projects with organisational strategy. She has a background in economics and planning and worked in market research and the transport sector before joining the MFB. Lyndsey also has extensive involvement in the development and monitoring of performance measures for the industry and has a personal passion for excellence in research method and in ensuring that research projects are written up so that others can learn from both successes and failures. Lyndsey is currently writing her own research thesis on research management as part of her Executive MBA.

INTERFACE BUSHFIRE COMMUNITY SAFETY

Painting by Mark Schaller

Chapter 1

Interface (urban–rural fringe) bushfire community safety

John Handmer and Katharine Haynes

The aim of this book is to present an accessible collection of bushfire research dedicated to community safety, being undertaken within the Australian Bushfire Co-operative Research Centre (CRC). The work is mostly undertaken by researchers from the social sciences and humanities, including economics and law. As far as we know, this is the first collection of research work on bushfire from these disciplines. The volume complements the very extensive published material on bushfires from ecology, fire behaviour, information management and related areas. It is an interim report on our research to date.

Our approach is to present the work in a way that is accessible to those responsible for bushfire risk management. This has guided our approach in obtaining feedback on each chapter from practitioners and in how the material is presented – as set out in the Acknowledgments. The research and review of the material for publication has been undertaken entirely in Australia. However, much of the material is generic in that it is likely to be relevant and useful to those dealing with bushfire community safety in other parts of the world.

We follow Australian practice in using the term 'bushfire', regarding it as synonymous with 'wildfire', 'forest fire' and so on as far as community safety is concerned.

History of the Bushfire CRC and development of the research themes

The Co-operative Research Centre (CRC) idea was developed in Australia in the early 1990s as a way to improve the uptake and application of research results. The idea is that bringing research providers and users together in one organisation means that research will be guided by the needs of users, who will have a stake in the research and therefore be more likely to adopt it. Most CRCs are commercial in nature, but the Bushfire CRC is largely a public-good corporation. With 34 agencies and research providers (universities and CSIRO) and over $100 million of resources during the funding term of 2003–10, both cash and in-kind, the Bushfire CRC is relatively large in CRC terms.

The research reported in this volume comes primarily from one of the CRC's four research programs – Program C: 'Self-sufficient communities'.

Even before the Bushfire CRC, fire research in Australia was quite well-known locally and internationally. It was undertaken mostly by CSIRO and some very well-known fire scientists, but uncertain funding and an ageing cohort appeared to be leading to declining research capacity. Although the fire and emergency sectors had some very large agencies there was little in the way of a research culture, dedicated research budgets were small and mechanisms for dissemination and adoption of new research were largely ad hoc. The Australian fire industry lacked a unified approach to research, although it was developing common approaches to issues through its relatively new industry body, the Australasian Fire Authorities Council (AFAC).

The bushfires in and around Sydney in the 2001–02 summer provided the impetus and political support, and AFAC commenced development of a CRC proposal. In addition to the substantive research program which had to be aligned to the needs of the industry, it was important to involve fire agencies and those with fire-related responsibilities. This was challenging, as the proposal had to be developed very rapidly in about nine months. A range of research providers including CSIRO, Bureau of Meteorology and universities were also needed. There were over 30 parties involved in the CRC application, which successfully went through a competitive selection process commencing operations in late 2003.

As part of the proposal, the research agenda was developed in close co-operation with end-user stakeholders from fire and land management agencies and other interested parties. Representatives from Program C were involved from the outset, and worked closely with Emergency Management Australia, the Bureau of Meteorology and some fire agencies. They also drew on the suggestions of researchers who contacted the CRC development group after national advertising, to develop a national research program in the community safety area. After an exhaustive process, the research agendas for the four research programs were finalised.

Program C: Self-sufficient communities

Our research in Program C and associated work elsewhere in the Bushfire CRC is dedicated to community safety and to the mobilisation of communities for their own safety. The research is:

- compiling the evidence to support efforts to reduce fire impacts through improving community safety and resilience
- developing tools and suggesting changes for improvements
- developing processes for assessing policy effectiveness and economic efficiency.

Details of the Bushfire CRC and research programs are on the CRC's website: www.bushfirecrc.com.

It is important to note that not all the bushfire research undertaken by contributors to this volume is presented here, and there is much relevant research outside the Bushfire CRC.

Book structure and themes

The book is organised around five themes:

- understanding communities
- assisting households and small businesses
- risk prevention and communication
- policy and institutional issues
- the future.

Here we very briefly highlight some of the issues covered in each section. These are indicative rather than comprehensive.

Understanding communities (Chapters 2–5)

An underlying philosophy in Australian bushfire management is that those at risk should play an active part in managing and reducing their own risk – and this approach is receiving increasing emphasis and acceptance. Implementation requires engagement with those at risk, development of appropriate policies and ways of assessing progress. It also demands ways of enhancing community self-sufficiency and resilience. In many communities this includes ensuring that community knowledge and capacities are acknowledged, utilised and strengthened. Making progress requires some knowledge of community expectations, perceptions, local knowledge and household strengths and weaknesses. An understanding of household decision-making dynamics – including the role of gender – with respect to fire emergencies may be key to the successful implementation of bushfire community safety.

Enhancing community members' capacity to come together in groups to develop their responses to the threat of fire can increase their resilience and, importantly, increase community confidence. The Community Fire Units developed in Sydney are examined in this context. Local knowledge and expertise is integral to these approaches. The role of local knowledge is discussed as a tool in fire planning and is explored as a crucial part of the community engagement process potentially leading to increased resilience.

Assisting households and small businesses (Chapters 6–8)

The cornerstone of community bushfire safety in most of Australia is the approach known as 'Prepare, stay and defend or leave early'. Essentially, the approach advocates that properly prepared householders should stay with their properties as a fire front passes – or leave the area well before the front arrives and evacuation becomes dangerous. This position moves away from the evacuation doctrine that prevailed among emergency services in recent decades (and still dominates overseas), towards greater community self-reliance. It is based on a number of important assumptions, including that houses can survive the passage of a fire front and will protect occupants from the fire, that occupants can be proactive in defending the house and save it at least long enough for the front to pass, and that last-minute evacuation is very dangerous and should be avoided. This is summed up in the catchphrase, 'Houses protect people and people protect houses' (AFAC 2001). It is also assumed that the legal context is supportive, although potential legal liability is always a concern among emergency service organisations.

Our research has compiled the evidence base and legal frameworks underpinning the policy, and work is continuing into implementation issues as well as examination of a comprehensive database on bushfire fatalities. Detailed research into how houses catch fire and burn down has shown that embers, rather than direct flame attack or radiated heat, are the main cause. This is consistent with residents being able to protect buildings from bushfires.

Risk prevention and communication (Chapters 9–11)

People can prevent bushfires, minimise fire-related loss and improve safety through appropriate behaviour. To do this, those at risk need to know how to reduce losses and enhance their personal safety, and everyone needs some knowledge about fire prevention.

Up to 60% of bushfires in Australia may be the result of arson, and many more result from carelessness. The remaining fires are caused by lightning. We can do little about natural processes such as lightning, but can have a major impact on human behaviour through education, training and incentives. This is not simply by formal processes. The media play a large, perhaps the major, role in informing the public about bushfires and community safety issues, even though fire agencies are very proactive in this area.

Relationships between the media and fire agencies are strong, based to some extent on mutual need. The complexities and different agendas of the two groups can sometimes lead to problems, an issue which receives attention in this section. Australian media generally take a very negative view of arson, especially when such fires lead to major losses. We know little about bushfire arson, partly because of limited data on the topic. Research is examining the motivations of arsonists and what strategies might reduce its incidence.

Improving levels of household preparedness is a priority for successful fire management and increased resilience. Aspects of research into the social and cognitive factors that support and undermine efforts at enhancing household and community preparedness are set out, along with steps that fire agencies can take to improve performance in these areas.

Policy and institutional issues (Chapters 12–15)

This part of the book deals with the fundamental issues of evaluation, economics and volunteer recruitment and retention. Evaluation takes place in the complex context of multiple definitions of community safety, highlighting the wide range of expectations in Australian fire and emergency management organisations that depend on volunteers, especially in interface and rural communities. Economics is a basic tool for assessing projects and programs – and provides support for decisions on resource allocation – but has not been widely applied in Australian bushfire management. Effort with economics research so far has concentrated on assessment of aerial firefighting, the value of volunteers, and valuation of all assets and flows including intangibles. An important challenge is to ensure that community safety is properly valued against the costs and benefits of fire suppression and prevention.

There is now a wide range of community programs being run by agencies across Australia, and the evidence suggests these are diversifying and occurring in greater numbers each year. To date, information about whether these programs are really working has been largely anecdotal and success has quite often been measured simply by quantity. Effective evaluation techniques are rare, especially in the context of multi-site public education and awareness programs. The questions of what worked for whom and in what contexts are examined using a program logic approach.

Challenges with volunteers include the need to invest in current members to retain their commitment, while making the organisations more appealing to new members and underrepresented groups such as females. An increasing problem is that of maintaining the relationship between volunteers and their employers to enable absences from work to attend fires and other emergencies.

The future with a warmer climate (Chapter 16)

There are many dimensions to the future, and many chapters deal in some way with this challenge. For example, Chapter 15 on volunteers is very much about dealing with the declining volunteer base in some areas, and the lack of participation by sectors of Australian society. The pressures on agencies to demonstrate that their approaches bring good value for money are discussed in Chapters 12 and 14. However, this section focuses rather more narrowly on climate and bushfires, as this is not covered elsewhere.

The climate has warmed significantly in Australia and globally over the last 50 years. The future appears almost certain to bring increased warming, and longer and hotter fire seasons with increased risk of major fires – often referred to as 'megafires'. The future is also bringing a rapidly expanding urban–rural interface. All major cities in Australia are experiencing rapid growth at the margins which are housing an increasing proportion of their inhabitants. The major underlying reason is probably economic, as outer-suburban housing is much cheaper than that in the city centre. It is also driven by lifestyle issues, including a desire to be nearer the bush – and often inadvertently at increased bushfire risk. This phenomenon of a growing interface is not limited to the edges of major cities. Those seeking lifestyle changes, and lower-priced property in many cases, are part of a massive development boom along much of Australia's coast ('sea-changers) and in some inland towns ('tree-changers'). Climate projections, demographic shifts and urban expansion lead to a much enhanced fire risk.

We can do little about the climate in the time frames we are concerned with, and once a major fire takes hold in weather conditions favourable for rapid fire development and spread, attempts at fire control become very expensive and of limited value. In Australia, preventing urban development is very difficult. Nevertheless, we can do much to reduce the impact and cost of fires through appropriate planning, building and landscaping, and through the attitudes and behaviours of those at risk. The future looks likely to present a stark choice – work on cost-effective protection of people and property or face escalating suppression costs and losses. Risk needs to be shared between those who bear the risk and the agencies charged with reducing it. Effective risk reduction involves a certain amount of action by the communities at risk as well as by the agencies. This book starts to set out the evidence base and guidelines for community protection.

Well-prepared communities, supported by fire services using imaginative approaches to informing people, can limit the losses from bushfires.

UNDERSTANDING COMMUNITIES

Painting by Mark Schaller

Chapter 2

Community perceptions of bushfire risk

Alison Cottrell, Sally Bushnell, Margaret Spillman, Judy Newton, David Lowe & Luke Balcombe

Chapter summary

The public often view and evaluate risk differently from researchers and experts. Understanding how the public construct their perceptions of risk can greatly improve risk communication, and direct risk reduction strategies most appropriately.

This chapter explores the social construction of risk in two peri-urban bushfire-prone communities in Queensland. These case studies were undertaken in 2005 using a multiple-methods approach, which included group interviews with community and fire brigade members, and a community survey.

While there are common factors that can similarly influence perceptions of bushfire risks within and between communities, there are often local-based issues unique to a community that have important implications for bushfire management. Through understanding and clarifying fire issues in communities, fire managers can address problems affecting bushfire risk mitigation in their local community. Engaging the community through a number of means could help considerably. The community should be viewed as a resource – communities have the capacity to act, despite vulnerabilities.

Introduction

The research reported in this chapter was undertaken to help facilitate the development of appropriate bushfire mitigation strategies at the local level by fire agencies, through developing a better understanding of the individual communities they serve. In particular, understanding how a community perceives a certain risk can greatly improve risk communication and direct risk reduction strategies most appropriately. Two case studies in peri-urban bushfire-prone communities in Queensland show how a picture of bushfire risk was constructed in the different communities. The case studies highlight the significance of locality – how factors at the local level influence community construction of risk, which can result in very different perceptions of the hazard within and between communities. This indicates that generic strategies to manage risk may not be appropriate in all situations, and a local focus may be required. This is problematic for fire agencies for two major reasons: first, they are not social researchers; second, they may not have the resources to adapt strategies at the local level. The final section of this chapter discusses ways to address this issue.

Social construction of risk

In contemporary social research, a useful approach for understanding disasters and hazards is 'social constructionism'. This kind of approach allows the acceptance of real environmental and social problems, but the way that society deals with these problems is socially constructed by the people who live in the particular society orculture (Spector & Kitsuse 1987; Holstein & Miller 1993; Hannigan 1995; Stallings 1995; Kreps & Drabek 1996; Lupton 1999; Oliver-Smith & Hoffman 1999; Hoffman & Oliver-Smith 2002; Lupton & Tulloch 2002).

Lupton (1999: 35) describes a number of approaches to understanding risk (Table 2.1). The realist position was taken in the research reported here. We accepted that there is an objective level of risk associated with hazards, which can be measured independently of social and cultural processes. However, public perceptions of risk can be moulded by social and cultural processes. Furthermore, it is important to understand how environmental knowledge, risks and problems are socially assembled, for example what is being said about the problem (claims), by whom (claims-makers) and how (process) (Best 1987; Hannigan 1995; Stallings 1995; Kreps & Drabek 1996).

Table 2.1   The continuum of epistemological approaches to risk in the social sciences

Epistemological position	Associated perspectives and theories	Key questions
Realist: Risk is an objective hazard, threat or danger that exists and can be measured independently of social and cultural processes, but may be distorted or biased through social and cultural frameworks of interpretation.	Technico-scientific perspectives/most cognitive science theories.	What risks exist? How should we manage them? How do people respond cognitively to risks?
Weak constructionist: Risk is an objective hazard, threat or danger that is inevitably mediated through social and cultural processes and can never be known in isolation from these processes.	'Risk society' perspectives/critical structuralism/some psychological approaches. 'Cultural/symbolic' perspectives/functional structuralism, psychoanalysis, phenomenology.	What is the relationship of risk to the structures and processes of late modernity? How is risk understood in different sociocultural contexts? Why are some dangers selected as risks and others not? How does risk operate as a symbolic boundary measure? What are the psychodynamics of our risk responses? What is the situated context of risk?
Strong constructionist: Nothing is a risk in itself – what we understand to be a 'risk' (or a hazard, threat or danger) is a product of historically, socially and politically contingent 'ways of seeing'.	'Governmentality' perspectives/post-structuralism.	How do the discourses and practices around risk operate in the construction of subjectivity and social life?

Source: Lupton (1999: 35).

This research aimed to identify community perceptions of bushfire risk, and how they differ from an objective level of bushfire risk determined by the Rural Fire Service (Qld). Social and cultural influences in each case study area helped to explain variations in observed community perceptions of bushfire risk. Fire service views on community perceptions of risk were also investigated, with the aim of identifying ways in which the perceptions of fire service providers and the communities they served differed and agreed on issues regarding bushfire hazards.

Community case studies

The Tamborine Mountain and Thuringowa case studies examined community perceptions of bushfire risks and its management, and bushfire service providers' perceptions of their community and hazard-related issues. Both studies used a multiple-methods approach. Multiple methods helps to reduce the bias associated with using only one method, and provides triangulation of the information. This approach is well established in the social sciences and leads to more confidence in research outputs (Denzin & Lincoln 1994; Neuman 1997; Hay 2005). Methods included group interviews and a community survey. Box 2.1 describes the group interview process, and Table 2.2 details survey delivery, response rates and respondent demographics. Full details of the methodology and results of the Tamborine Mountain and Thuringowa case studies are available in Balcombe (2007) and Bushnell et al. (2006) respectively.

Figure 2.1   Bushfire researcher Jenny Indian talks to Ron Leary and Paul Chambeyron, members of the public. © Bushfire CRC.

Box 2.1: Summary of group interview process

Group interviews

A number of initial interviews with key brigade personal and community members broadly defined bushfires issues in the study areas, and directed question development for group interviews. Local brigades and community organisations were then invited to participate in group interviews, which were held during February 2005 in Tamborine Mountain and between April and July 2005 in Thuringowa. The group interviews helped define local bushfire issues from the brigade and community group perspectives, and contributed to the development of the community surveys.

Table 2.2 Community survey data collection and respondent details

Community survey	Tamborine Mountain	Thuringowa
Data collection	500 surveys delivered to letterboxes (May 2005), based on representative sampling methodology: 2218 households in survey area = 1 survey to every 5th house. Postage-paid envelope provided for survey return.	947 surveys delivered to letterboxes (October 2005), based on representative sampling methodology: 100 surveys delivered in each of 9 RFB areas similar to Tamborine; 57 delivered to every household in 1 RFB area (due to small population). Postage-paid envelope provided for survey return.
Response rate	163 surveys returned (33%)	263 surveys returned (28%)
Gender	Male (53%); female (47%)	Male (46%); female (54%)
Age	18–24 years (1%) 26–40 (10%) 41–55 (30%) > 55 (59%)	18–24 years (2%) 26–40 (22%) 41–55 (34%) 56–70 (32%) > 70 (9%)
Primary household types	Couple with no children (33%) Couple with children left home (21%)	Couple with no children (26%) Couple with children left home (23%)
Primary occupations	Professional management (35%) Self-employed (21%) Tradesperson/skilled worker (13%)	Professional management (28%) Tradesperson/skilled worker (19%) Officer worker/white collar (15%).
Primary employment status	Full-time (31%) Retired (45%)	Full-time (47%) Retired (28%)
Primary property types	Residential street (61%) Escarpment (19%) Bushland/forest (13%)	Rural (67%) Suburban (24%) Farming (7%)
Primary length of residency at current address	1–5 years (44%) 6–10 years (18%) 11–15 years (17%)	1–5 years (34%) 6–10 years (19%) 11–15 years (19%)
Home ownership	Own home (94%) Rent (6%)	Own home (90%) Rent (10%)

Tamborine Mountain case study

Tamborine Mountain is located in the hinterlands near Brisbane, Queensland. The bushfire risk in Tamborine has been rated as high by fire services (RFS 2002a), however, survey respondents generally underestimated the risk. The case study identified a number of factors that may explain this observation. Tamborine Mountain is a very picturesque area and the survey showed that amenity values are very important to people living there. This benefit of living in Tamborine Mountain may have moderated risk perceptions. This issue was raised by community members in the group meetings: residents living on the edge of escarpments understand their position increases their risk, but the view is given priority over the danger. Gilbert (2004) and Collins (2005) similarly noted the effect of non-hazard benefits on community risk perceptions, and McCaffrey (2004) explained that people tend to balance the perceived risk and benefit of living in a particular area – the higher the perceived benefit, the greater the risk tolerance.

Undertaking bushfire preparation activities has been identified as a factor which can modify risk perceptions (Montz 1993). Taking steps to mitigate the impact of bushfire can indeed reduce risk, so homeowners doing so thereby view themselves as less vulnerable. However, unless the whole suite of strategies needed to safely protect a property during a bushfire have actually been undertaken, the homeowner has not significantly reduced their vulnerability. This possible mix of underestimating risk and overconfidence in facing a risk may in fact increase vulnerability. Many respondents in Tamborine reported being confident in many aspects of bushfire safety. For example, 75% of respondents said they were confident in preparing their homes to minimise the impact of bushfire. However, most respondents had not undertaken a number of crucial bushfire protection activities. In particular, more than half had not decided when they would stay and defend their homes during a bushfire or decide to leave early. An extensive firebreak network in the area, created as a community project, may instil confidence in residents, which may or may not be founded. The lack of reticulated water in Tamborine Mountain means that residents rely on water tanks, an important step in protecting homes against bushfire. However, interviews with community groups indicated that residents often do not store enough water even for personal use during the dry season, and therefore do not have enough to defend their homes during a bushfire.

Tamborine Mountain is a particularly dynamic area in terms of community structure. It is close to Brisbane and has seen an influx of people from the city, both retirees and commuters, and is a popular tourist area. Such a situation has major implications for the construction of risk in a community. Newcomers to an area cannot identify their level of exposure to risk until they assess the hazards to which they are exposed (Beringer 2000). There may be a significant proportion of the Tamborine Mountain population (newcomers and tourists) who are unaware of the bushfire risk. Community groups were concerned that many people may not recognise the risk because the area is often green. They also raised the issue of tourism, and the need for assistance in developing bushfire plans with tourism operators.

The survey indicates that Tamborine Mountain residents are concerned about many risks other than bushfire. For example, most agreed that they were less concerned about the risk of fire than about other risks to personal safety, and most disagreed that the impact of fire is far greater than that of any other risk. Personal and family matters, road safety, crime and storms were listed as personally more important than the bushfire hazard. Community groups also indicated that people were concerned about development and environmental issues. Fire brigades in Tamborine Mountain attend more road accidents than bushfires, which objectively indicates that road safety may be a greater risk to residents. Covello and Johnson (1987) explained that the array of risks to which people are potentially exposed is so vast that it is impossible for individuals to be aware of all risks. Therefore 'people must decide which risks to fear most, which risks are worthy of attention and concern, which risks are worth taking, and which risks can be ignored' (p. viii). The risk selection process can be without scientific evidence, does not necessarily relate to real danger and may include risks which are among the least likely to affect people; it can be largely determined by social and cultural factors, not nature. Given that the community at Tamborine Mountain is especially dynamic and hence the social and cultural influences are complex, including the mix of rural and urban characteristics, it is not surprising that bushfire hazards compete with other risks for attention.

Community cohesion has also been found to affect risk perceptions – a close-knit and structured community tends to better perceive natural hazard risks (Gough 2000; Flint & Luloff 2005). Although it could not be confirmed from the survey data, in such a dynamic and large community as Tamborine Mountain it is likely that the community is not cohesive as a whole, which can also explain the observed tendency to underestimate the bushfire risk.

Group interviews with the fire brigades in Tamborine Mountain suggested that they have a good overall understanding of their community's perceptions of bushfire risk and related issues. Brigades identified a general lack of awareness or acknowledgment of bushfire hazards. They also identified vulnerable groups including the elderly, and residents who compromise safety for aesthetics. However, they may be underestimating the capacity of their community. While there was acknowledgment of strong community views and passionate groups relating to environmental and bushfire issues, there appears to be a view that the community is more or less dependent. For example, it was mentioned that the community expects too much from fire services. Results from the survey indicate that residents view themselves as primarily responsible for protecting their property, although they did expect the brigades to be able to provide clear advice. Other studies have also shown that residents accept some degree of responsibility for risk reduction (e.g. Winter & Fried 2000).

Thuringowa case study

Thuringowa is located in north Queensland and is the twin city to Townsville. Bushfire risk in the Thuringowa area is generally less than that of Tamborine Mountain, being predominately rated as moderate (RFS 2002b). Thuringowa respondents appeared to perceive this risk accurately – the majority rated the hazard in their locality as moderate to high, and most also rated their concern about the bushfire hazard as high. Thuringowa is less renowned than Tamborine for picturesque landscapes and respondents nominated a variety of reasons, aside from a rural lifestyle, for moving to the area, such as family or a life change including retirement. Although these are benefits, which may modify risk perceptions, they are not directly related to features that promote bushfire risk, such as the forests in Tamborine. In fact, Thuringowa respondents indicated that peace, quiet and space were the features that they valued most, above trees and bushland. Without the need to balance perceived risks and benefits, Thuringowa residents may be better able to perceive the bushfire risk in their area.

Perceptions of personal risk, however, may be inaccurate. Respondents consistently rated the risk of bushfire at the household level lower than that of their locality. This was not investigated in Tamborine, although it may be a common phenomenon – people tend to view themselves as less at risk than others. For example, Odgers (2002) found that even with adequate threat knowledge and risk perception people do not always perceive their own susceptibility to the threat, and often consider that in the event of a hazard they would be resilient and recover quickly. This may be the result of normalisation bias: those who have experienced a hazard in the past may infer from their ability to cope in that particular instance that they will be able to cope with any future occurrence of the same hazard (Johnston et al. 1999; Kumagai et al. 2004). It may also be the result of optimistic bias: people can simply rate themselves less vulnerable and more skilful than average (Johnston et al. 1999; Kumagai et al. 2004), or deny the risk (e.g. 'it won't happen to me') (McCaffrey 2004).

In terms of bushfire preparation, interviews with community groups revealed a common theme: complacency. Interviewees were somewhat divided on whether residents understood the bushfire risk but agreed that many people were complacent or indifferent in terms of reducing the risk of bushfire. The most obvious explanation from the survey and interview data is that the cyclone hazard undoubtedly takes precedence. Cyclones are a seasonal hazard in the area and, unlike bushfire, past impacts from cyclone events have been significant and well documented. Consequently, perceptions of catastrophic potential and feelings of dread are felt by locals. These are major factors leading individuals and communities to evaluate the hazard as serious (Covello & Johnston 1987). How this affects bushfire preparation was not measured in this case study, but it is likely that preparation for cyclones also takes precedence.

As discussed previously, sometimes people select some risks and not others. In Thuringowa, where a person may not be able to address a multi-hazard situation because of resource or time constraints, bushfire hazards may be disregarded in order to address the hazard considered to be more important. The householder may be viewed as complacent or apathetic, as described by an interviewee.

One community group, however, identified affordability and access to equipment as a problem. Access to resources is an issue for many people in many settings. In Forest Ranch (California), Collins (2005) identified household access to resources as a primary determinant of hazard vulnerability. In Thuringowa, properties with high fuel loads and people dumping household and green waste in inappropriate places were major issues raised in community and fire brigade group interviews. This may indeed be the result of indifference, but the survey revealed that there might be problems with waste-disposal facilities. Many respondents felt that accessibility and the cost of waste-disposal facilities were deterrents. Other cost and access to resource issues were evident in survey data from both Thuringowa and Tamborine. Therefore, while a lack of action to reduce bushfire risks by those who correctly perceive the risk may be the result of indifference, it is important to investigate further.

The survey area in Thuringowa is a peri-urban area and, similar to Tamborine, the community is characteristically dynamic. Community group interviewees clearly viewed the influx of newcomers, particularly those from urban areas, as a major issue. Newcomers were identified as a group who did not understand the bushfire risk and did not participate in risk mitigation strategies. These associations were not significant in the survey data, but as other studies have reported significant findings (e.g. Beringer 2000; McGee & Russell 2003) further investigation may be warranted. Typical of peri-urban areas, the Thuringowa community did not appear to be cohesive. A lack of community cohesion may also help explain the lack of bushfire preparation in both surveyed communities: closer-knit communities tend to be more resilient' (Anderson-Berry 2003: 226).

Group interviews with fire brigade groups in Thuringowa revealed a good understanding of fire issues in their community. In particular, brigades had suspected a link between waste dumping and access to waste-disposal facilities. Lack of community participation in risk mitigation was a major theme and newcomers, particularly those from urban areas, were identified as a group who did not prepare. The overall view was that the community had an attitude of indifference and expected the brigades to take responsibility for all risk mitigation. As at Tamborine, the survey data did not support this view. The community was not necessarily ignorant of the risk, rather they had a relatively good understanding and viewed themselves as responsible for bushfire risk reduction on their own property. Perceiving the risk and personal responsibility for reducing the risk should lead to better preparedness (Beringer 2000; Winter & Fried 2000; Odgers & Rhodes 2002; McCaffrey 2004), but because residents overall are not adequately prepared there may be issues within the community that are not well understood by brigades.

Understanding community risk perceptions

Evaluating risk: experts and the public

'Risk' is ambiguous, it carries both probability and consequence: risk increases as the probability of a negative event increases and as the expected consequences grow worse (Sjoberg 1999). Experts tend to stress the probability of a risk, while the public judge a risk based on a complex set of characteristics that incorporates consequences rather than probabilities (Sjoberg 1999; Renn 2003). It is difficult for the public to understand the concept of probability because it often incorporates theoretical models, which require a large precision when dealing with rare events (small probabilities) that have large consequences (Sjoberg 1999). It is also difficult for the public to evaluate a risk without considering other potential risks and personal matters. For risk experts, it is difficult to understand and accept that risks are also socially constructed, and thus perceived differently. However, as public perceptions often drive priorities on where and how to reduce or manage risks, adequately understanding perceptions is crucial to developing good policy (Renn 2003; Byrd & VanDerslice 1996).

The need for a local-based approach

The way in which the public perceives a risk is highly complex. Perceptions are shaped by social and cultural processes as well as psychological factors, in addition to some degree of reference to scientific evidence. As demonstrated by the case studies, risk perception is often unique to different communities and to different individuals. The implication for hazard management is that locality is important. Although different localities share similar issues, there are local-based issues that need to be addressed to ensure effective hazard management.

Tamborine and Thuringowa are peri-urban areas and characteristically share issues related to an influx of new residents, who often do not understand the fire risk and are thus not likely to be prepared. Therefore, common strategies could reduce the vulnerability of this group. The Tamborine and Thuringowa communities differed in their perceptions of bushfire hazards in their locality, and there were important local-based issues that affected bushfire hazard management. For example, risk perceptions and bushfire preparation in Tamborine appeared to be moderated by the perceived benefits of living in a picturesque forested area. Thuringowa residents appeared to perceive the bushfire hazard risk level accurately, but that did not necessarily lead to household preparedness. Residents in Thuringowa clearly perceived a more serious risk: cyclones, which may hinder bushfire preparations. Furthermore, the survey indicated that inadequate waste-disposal facilities may be preventing many residents from managing vegetation on their property, thereby increasing the risk of bushfire. Importantly, both case studies revealed that some residents in both localities have problems accessing resources, and are therefore limited in their ability to reduce their level of risk.

Community involvement in hazard management

Byrd and VanDerslice (1996) compared perceptions of environmental risk in three communities in El Paso (Texas) and found important differences relating to local issues. They linked vulnerability with a lack of agency understanding of the community it serves. In Tamborine and Thuringowa, the case studies highlighted a number of points: potential issues relevant to bushfire management were not recognised by the brigades; there were issues that the brigades were aware of but the source or extent could not be accurately confirmed; and the brigades have a good understanding of a number of community fire issues. It is therefore important to actively investigate fire issues in communities to clarify which issues are understood and which are not well understood. This should lead to better hazard management. Consulting the community and involving them in risk decision-making can help considerably in highlighting bushfire issues and defining courses of action (Box 2.2). From the community's perspective, it is important that their views be heard (and valued) and that their perceived right to have a say in developments that affect their lives is fulfilled.

Box 2.2: How agencies can benefit from community consultation

Benefits of community consultation in hazard management

- Informing and educating the community about issues
- Tapping into community knowledge and possible solutions
- Understanding community preferences for hazard management equity
- Achieving practical and effective outcomes (Cottrell, 2005).

There have been some successes in involving communities in hazard mitigation, although often accompanied by criticism. The growing amount of literature on the subject is providing evidence and practical solutions for positive outcomes. Burby (2001) lists a number of choices for hazard managers that are relevant to the efficacy of community involvement (Box 2.3). Cottrell (2005) also demonstrated the importance of extensive and early community participation, and emphasised that the contribution of participants in the broader community should not be underestimated.

Box 2.3: Community involvement in hazard management

Promoting positive outcomes from community involvement

Burby (2001) suggested that hazard managers consider the following.
- Establish clear objectives
- Time participation so that it reinforces the chosen objectives
- Involve all stakeholders

In bushfire management at the local level, particularly where rural volunteer brigades manage the bushfire hazard, engaging the community is a challenge with the limited resources available. However, to ensure efficient and effective bushfire management – which is especially important when there are limited resources – engaging the community should be a high priority. Therefore, techniques for doing so are required.

Ways forward

The case studies in this chapter highlighted similarities and differences between the community and their brigades, and between the community and community group representatives. There were many issues that the community groups were able to articulate, that is, community group representatives' perceptions of fire issues in their community closely matched those identified by the surveys. Brigades may be able to gain a better understanding of their community and issues related to bushfire management by engaging key community representatives. This needs to be tested further. Future case studies could test the robustness of the case studies presented here. Investigating the similarities and differences between community groups and the community that they represent in terms of demographics, using census data for example, is also important. An output of this research would be a tool for fire managers, applicable at multiple levels of management including the local fire brigade level, to address fire issues locally.

Conclusion

Resources available at the local level in particular will remain a challenge for fire management. This is why fire management often becomes the responsibility of the local community, who are not always equipped to fulfil that responsibility and are often not even aware of it. This, among other reasons discussed, leads to a lack of community participation in hazard management, and often a view that the community is not interested or capable. Flint and Luloff (2005) argued that despite vulnerabilities all communities have the capacity to act. The Tamborine Mountain and Thuringowa case studies, as well as the study by Byrd and VanDerslice (1996), indicate that despite an overall lack of preparedness, communities recognise that they are responsible for hazard management to a degree and they believe that the risk can be managed. However, importantly, they require some direction – clear advice on actions they should take. Communities should therefore be viewed as a resource, and tapping into this resource, through engagement and building its capacity to act, has the potential to address fire management with limited funds.

The Tamborine Mountain Escarpment Management Strategy (Watson 2001) provides a good example of community capacity when adequately supported. Initiated by members of the community, the strategy was developed with various groups including representatives from the relevant councils, Queensland Parks and Wildlife Service and the Rural Fire Service. Funding through the National Heritage Trust supported the development of a comprehensive strategy that, in addition to addressing on-ground management of bushfire hazards, included avenues for community education and involvement in reducing bushfire risks. With continued support and ongoing implementation of strategy recommendations, residents' tendency to underestimate the bushfire risk in Tamborine Mountain may be addressed and the bushfire hazard better managed on both public and private lands. Other communities, including Thuringowa, have the potential to develop similar strategies with appropriate support. The case study undertaken in Thuringowa highlighted a number of local issues that could be addressed by a bushfire hazard management strategy, which could significantly improve the safety and resilience of residents.

Practical outcomes and recommendations

- Local-based issues can lead to varying public perceptions of bushfire risk within and between communities, which has important implications for hazard management.
- Local fire issues can be better understood and clarified through community engagement.
- Communities have the capacity to respond to the bushfire threat. The majority of respondents accepted responsibility for bushfire hazard reduction on their properties, but indicated that they require support such as clear advice.
- Communities should be viewed as a resource. With appropriate support, community and collaborative activities can address bushfire hazard reduction.
- Further community-based research, with the aim of developing and refining tools that allow fire managers to better engage their communities particularly where resources are limited, is recommended.

Chapter 3

Resilience at the urban interface
The Community Fire Unit approach

Tom Lowe, Katharine Haynes & Gerry Byrne

Chapter summary

Community Fire Unit (CFU) initiatives in New South Wales and the Australian Capital Territory are increasing in popularity and cost. This unprecedented approach requires detailed analysis in order to quantify its effectiveness in achieving key community bushfire safety goals and to gain an idea of the likely challenges as numbers grow and new issues arise.

An assessment was carried out utilising a range of qualitative and quantitative research techniques, the breadth of which was intended to provide a robust and objective set of results from which key issues could be identified and discussed. Data have been compiled from 20 agency interviews, 670 CFU member questionnaires, 50 public questionnaires and four focus groups. The issues discussed in this chapter are led primarily by the findings of the survey and focus groups.

The research identified an overall sense of empowerment, increased capability and social capital among individuals involved in the CFU program. The training and equipment were particularly valued as they boosted confidence and created more of an active and co-operative role for groups intending to stay and defend their homes from bushfire. However, it was also found that a minority of CFU members tended to focus too sharply on their operational role, leaving gaps in their wider preparation and planning. In addition, the strong community links within the CFU movement were identified as having a potentially negative influence on wider community relations, preparedness and communications.

It is hoped that this research will not only advance the CFU program but also be relevant to everyone involved with promoting community resilience.

Introduction

The vulnerability of urban interface communities to bushfire is at an all-time high. Reductions in prescribed burns, the consequent build-up of fuel loads and an increasing desire to live close to the bush cause problems for fire authorities. Add to this predictions of longer and drier fire seasons in the future and it becomes clear that high-consequence events are likely to become more frequent and more intense (Pitman et al. in press). In a situation in which environmental, social and physical factors conspire to seriously threaten lives and property, such as in Canberra in 2003, current resources and levels of preparedness in interface areas are clearly not enough to cope with the demands created by such events.

Since the first UN World Conference on Natural Disaster Reduction (WCNDR) there has been a drive towards increased preparation and resilience rather than response and recovery. This was reiterated at the WCNDR in January 2005 in the Hyogo guidelines for disaster-resilient communities, which stated that negotiation and bottom-up initiatives should be used to achieve a more effective community response and greater resilience. This has been noted by fire authorities across Australia, encouraged by a federal commitment to research, contingency planning and policy reform. A major part of this reform has been the restoration of community responsibility for their own safety and survival with regard to certain natural hazards. As a result, disaster management in Australia has shifted from response and recovery to risk reduction, a shift that has realised the need for public participation (Pearce 2003). Pearce (ibid) noted that while a strong top-down policy is needed, local-level bottom-up policy is required to implement mitigation strategies and manage disasters in an effective way.

By identifying the need to create partnerships with at-risk communities, emergency management authorities have broadened and reduced their own burden of risk. However, the development of these partnerships is problematic (Chess et al. 1995) with the notion of 'resilience' assuming trust, shared responsibility, co-operation and communication across numerous organisations, professionals and community groups (Bullock & Haddow 2003). In addition, educating and empowering communities in order to successfully mitigate the harmful effects of bushfires extends far beyond people knowing what to expect and how to react when threatened (see Ch. 11 this volume). Thus, active public involvement is needed if fire agencies are to successfully harness local expertise and knowledge and create a sense of ownership and community responsibility among those at risk. Pursuing these goals is a worthy activity – despite the difficulties, there is clear evidence that, through collective learning, groups can address risk effectively even though levels of understanding, commitment and skill may vary (Nilson 1995; Comfort 1999).

Importantly for this study, the 'Prepare, stay and defend or leave early' bushfire strategy can only truly be effective if it is preceded by adequate preparation (Gledhill 2003). This requires the public to go beyond simple practical actions such as creating a defendable space. It involves trust in and access to information, individual psychological preparedness and effective interactions between the public and fire authorities. Developing this kind of preparedness, particularly in urban areas with low risk awareness, limited bushfire experience and weak social interaction (Cottrell 2005), is a challenge that has yet to find a robust solution.

The urban interface: a developing hazard

The urban interface is a complicated environment for fire agencies to successfully operate in. Challenges include the sheer volume of properties at risk, the dynamic nature of fires, lack of community awareness and education, large-scale self-evacuation, narrow streets, failure of essential services such as gas and electricity, and inter-agency communications. In the traditional bushfire model, fires are started by lightning strikes in rural locations. The development of large-scale fires often results from adverse weather conditions and large volumes of combustible vegetation, causing the fires to make runs that threaten urban communities in their path (Bradstock et al. 1998). In this scenario, there is time to plan and develop mitigation strategies prior to any impact on the urban interface. In the urban interface, however, a new paradigm is now evident – the short, sharp interface fire. Lasting no more than one or two days these fires develop swiftly and can affect numerous properties simultaneously. Rohde (2002), in an assessment of urban interface fires in California, noted that 'Contemporary wildfires in urban interface areas are forming new patterns. Urban interface fires, unlike their 20th century counterparts, require neither protracted time nor large acreage to become destructive' (p. 33). He also stated that 'Given the ever increasing population and movement of new communities into former wildlands in California, this kind of fire has the potential to become the dominant high-loss wildland structural interface fire of the future' (p. 221). The similarities between California and New South Wales are evident not only in the types of fires they are experiencing, but in the continuous expansion of communities into previously rural areas. These urban interface fires develop quickly, and have the potential to quickly overrun local firefighting resources. In this environment it becomes essential to develop specific and appropriate community resilience to these fast-moving events.

The Community Fire Unit approach

The NSW Fire Brigades Community Fire Unit (CFU) approach was initiated in 1994 following calls for greater community control over lives and property, homeowners' expressions of willingness to stay and defend and a realisation that the number of structures at risk cannot be successfully defended by fire services alone. The response has been a commitment to prepare communities at a street level to carry out this task in predominantly urban interface areas.

Differences between rural communities and their urban counterparts and the type of volunteers who serve them dictated the development of CFUs. The urban interface appears to have generally less sense of community responsibility than rural areas, and a greater reliance on agency or government support in times of crisis. The socio-economic reality in these areas, i.e. many dual-income families of a professional or semi-professional nature with busy lifestyles, makes a minimal commitment very attractive and is largely responsible for the success of the program. The program also appeals to volunteers' sense of self-interest, dictated by its territorial nature. The NSW Fire Brigades, recognising these differences, felt that the traditional rural volunteer model would not work effectively in the specific context of the urban interface. Instead, they offered urban interface communities a model that would provide a level of self-reliance with a minimum of commitment.

The program can best be described as a hybrid volunteer system; developed to meet the specific needs of the target communities. CFUs have no response role whatsoever and are not provided with fire trucks. They are specifically tasked with protecting their properties from ember attack and spot fires and protect a designated geographic area, usually a street or part thereof. CFU members do not have the same level of training, commitment or responsibility or the broad scope of Rural Fire Service or State Emergency Service volunteers. Nor do they share the autonomy and freedom in decision-making encouraged in the majority of community-based education and preparedness schemes, e.g. Community Fireguard in Victoria, Community Fire Safe in South Australia and Street Fire Wise in New South Wales. Current NSW Fire Brigades practice is to canvass communities that are regarded as being at risk. However, the application process for the commissioning of CFUs depends upon community expressions of interest.

By combining top-down or 'command and control' management (CFUs are covered by the *Fire Brigades Act 1989*) with bottom-up proactive community involvement, NSW Fire Brigades have formed a unique voluntary wing. The running of this movement must balance strong leadership and strict hierarchy with the people management skills required to maintain volunteer preparedness for infrequent but potentially devastating events. Membership offers CFU volunteers protective clothing, basic firefighting equipment and training in return for a minimum number of hours training to meet operational requirements. CFU volunteers are not required to protect any other properties than those in their immediate surrounds. In this respect there is some commonality with the original bushfire volunteers who were dedicated to and worked in their local community only.

The CFU approach is intended to empower community members to be proactive in the defence of their own properties by utilising existing social capital and local knowledge (see Ch. 4 this volume) as an important resource. This capital and information is enhanced by providing communities with the equipment, further knowledge and improved social networks required to carry out a limited but important role. It is hoped that this fusion of expert and local skills, knowledge and networks will produce a more fire-resilient urban interface.

While CFUs can create much-needed flexibility in urban fire brigade resources during major bushfire events, a secondary effect relates to implementation of the 'Prepare, stay and defend or leave early' policy, the central tenet of the Bushfire CRC program C6 (Ch. 8 this volume; Handmer & Tibbits 2005). The concept depends upon public preparedness. In the case of CFUs, this preparedness is provided in a single package, the content of which is hopefully enough to save homes and lives. The success of the CFU approach was demonstrated when a bushfire hit the urban interface at Cross St, Warrimoo, in the Blue Mountains during the 2001 fire season. Cross St CFU members successfully implemented mitigation strategies to protect their properties in line with the training provided by the NSW Fire Brigades.

In terms of interest and numbers, the CFU scheme has been a great success (Figure 3.1), with the red and white livery of the NSW Fire Brigades a familiar sight on trailers and hose-posts throughout interface areas of Greater Sydney and the Blue Mountains area (Box 3.1). A version of the scheme has also been implemented in the ACT. This success and growing membership runs counter to declining volunteer trends elsewhere in the emergency management field (Ch. 15 this volume).

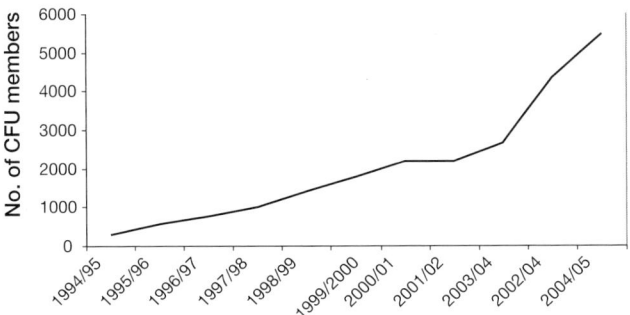

Figure 3.1   NSW Fire Brigades CFU membership, 1994–2005
Source: NSW Fire Brigades

Given the scheme's growth in size and popularity it is deemed important that any future development should be carried out with a good idea of its advantages and possible problems. Thus, using data gathered from focus groups and a CFU member survey, this chapter will discuss the ways in which this brand of community empowerment acts to promote bushfire awareness, preparedness, positive behaviour change and ultimately resilience in urban interface areas. It suggests where this kind of scheme may fit into bushfire management in the future and will interest those working in community outreach and education roles, principally within fire organisations. Rhodes and Gilbert (Ch. 12 this volume), use an alternative set of research questions and methodology to investigate the CFU program.

In particular, the chapter addresses a number of research questions.
- Understanding the social dynamics of CFUs: what are the attributes of people who becomes involved? What kinds of communities are likely to have CFUs and why?
- Understanding the effect of CFUs upon communities: how have people benefited from their membership? Has involvement promoted community resilience to bushfires?

- Understanding the effect of CFUs on perceptions of fire risk and household behaviour in fire-prone communities: What impacts do these effects have upon decisions to leave early or stay and defend property?
- Understanding how to manage the growth and development of the CFU movement: what issues need to be addressed? How should problems be addressed?

Box 3.1: NSW Fire Brigade's CFUs

Program started: 1994
Number of units: 330*
Number of volunteers: 6000*
Average members per unit: 15
Unit structure/hierarchy
- NSW Fire Brigades officer (overseer)
- CFU leader
- CFU custodian
- CFU regular members

Equipment provided
- Personal protective equipment
- Hoses
- Hose fittings
- Davey pump

Equipment cost: $15 000 approx*
Official training provided: min 12 hours/year (optional extra unsupervised training)
Overall program cost: $2.6 million*
What do CFU members learn about?
- Bush care and bushfire behaviour
- Safe housekeeping and gardening practices
- Planning and preparing for bushfires
- Operating and handling firefighting equipment
- Mopping-up operations
- Processes that help to reduce bushfires in the community
- Limiting the effects of bushfires on life, property and community in times of bushfire.

*Figures accurate for May 2007.

Methodology

In designing an appropriate methodology for this assessment, we considered a number of factors. First, the main research question relates to implementation of the 'Prepare, stay and defend or leave early' policy at the urban interface. To a certain extent, the success with which responsibility for bushfire safety can be handed over to the individual by fire authorities depends on existing levels of awareness or perceived risk. Risk and vulnerability research have shown that a number of factors influence the likelihood and ability of individuals or communities to act to reduce their risks. These include previous experience, gender, cultural, social and economic background and political representation (Slovic 2000; Wisner et al. 2004; Haynes et al. in press). Information of this sort was considered important as it would, for example, indicate whether those most likely to be involved in CFUs considered themselves at greater risk than people who were less inclined to become involved.

Second, it was felt that within the strongly top-down structure of the NSW Fire Brigades there was scope for ideological differences between the corporate strategists, the administrative/ implementation level and the CFU members. In addition, the developing nature of the CFU movement necessitated a state-of-the-art in terms of its current position and intended direction. There was also a need to gather views on the CFU approach from a wide range of informants involved in community education and preparation and therefore avoid a biased sample.

CFU member questionnaire

It became clear early on that even basic demographic information on CFU members did not exist, a situation which suggested the need for a member survey. Due to the wide geographic spread of CFUs and their marked reliance upon email, an on-line questionnaire was most likely to reach and be completed by the large numbers required to make the survey accurate. It would also reduce the time and expense of other survey methods. A small number (n = 40) of face-to-face surveys were carried out at a training day to collect data from people who may not have access to email. A total of 670 questionnaires were completed.

The survey, developed through a wide range of input and pretesting, contained 30 open and closed questions, starting with demographic information, i.e. age, gender, occupation, education, income, level of insurance and number of children. Next, respondents were asked about their experience of bushfires and the extent to which they had been personally affected by bushfires. They were then asked to consider how likely it was that they would be affected by bushfires on a range of timescales. Respondents were asked to indicate (using a 5-point scale) the degree to which they agreed or disagreed with a number of statements outlining motivations for becoming a CFU member. Finally, they were asked to rate their preparedness and the preparedness and knowledge of non-CFU members in their area. Several opportunities were provided for open-ended comment.

The URL internet link to the survey was emailed to CFU members along with a covering letter using the NSW Fire Brigade's email list. It is acknowledged that this list was not exhaustive and that some members may not have had internet access, but it was felt that the method nevertheless targeted a wide range of geographic and socio-economic groups.

CFU member focus groups

Focus groups were used to further develop understanding of the preliminary survey and agency interview findings,[1] focusing on the more qualitative aspects of individual risk perceptions and motivations. In addition, they provided an opportunity for CFU members to introduce various issues and discuss them in depth. Four focus groups were undertaken, each comprising seven or eight CFU members. The two-hour meetings were spread throughout the Sydney area (North Rocks, Heathcote, South Turramurra and Glenbrook (Blue Mountains)) to represent a range of geographical, demographic, socio-economic and experiential factors. Locations were selected by NSW Fire Brigades administrative staff, based upon the requirement for a diverse sample. Discussions were led by a facilitator using a generic protocol, however, they were allowed to digress if the content was deemed relevant.

Overview of survey findings

Survey findings are presented in Tables 3.1 to 3.8, and in Figures 3.2 and 3.3.

Table 3.1   CFU survey figures

Number of surveys completed	670
Percentage of CFU volunteer community surveyed	11.6%

Table 3.2   CFU member age range

Age	%
<25	2.7
25–35	2.9
36–45	21.6
46–55	35.9
56–65	25.9
>65	11.1

Table 3.3   Gender ratio

Gender	%
Male	78.9
Female	21.1

Table 3.4   Home ownership and insurance

Home owners	92%
Home and contents insurance	97%

Table 3.5   CFU activation

Have had their fire unit activated	43%
Have never been activated	57%

Table 3.6   Top three motivations to join a CFU

Motivation	%
'I know that I live in a bushfire-prone area so I wanted to be able to protect myself/property/family in the future'	36
'Bushfires affected my local area in the past so I wanted to be able to protect myself/property/family in the future'	22
'If I and my neighbours are trained and equipped to defend our homes it gives us the greatest chance of saving them'	20

Table 3.7   Perceived level of success of information transfer to the wider community

Unsuccessful	43%
Somewhat successful	45%
Successful	12%

Table 3.8   Reasons for CFU member drop-out (from existing members)

Reason for leaving CFU	%
Moved from the area	53
Other commitments took precedent, e.g. job, family, other voluntary groups	26
Became uninterested	18
Became ill/passed away	13
Disagreement within the group	4
Dissatisfied with NSW Fire Brigades management of CFUs	3
Too concerned about the dangers involved	1
Other	8

* Percentages add to more than 100% as respondents nominated more than one reason.

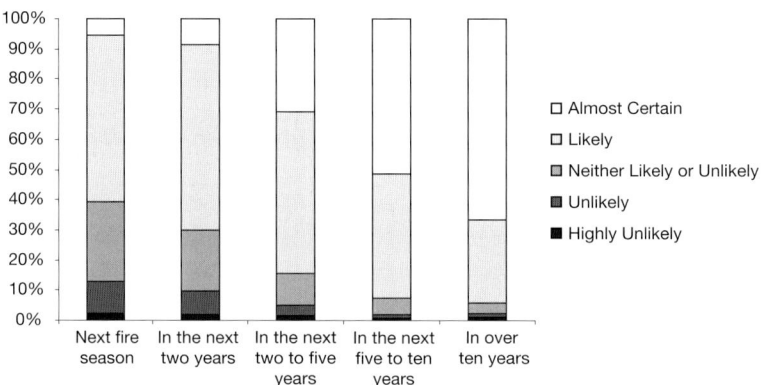

Figure 3.2   Likelihood of bushfire in local area over time

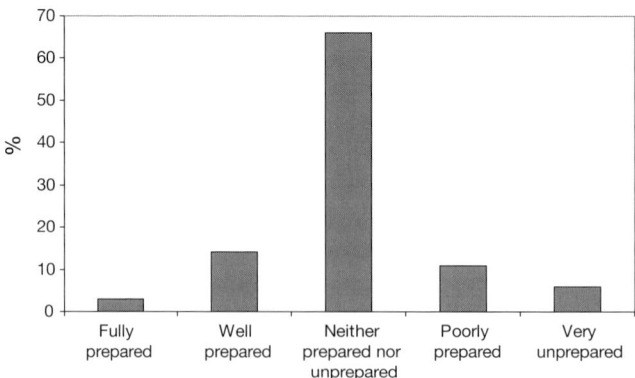

Figure 3.3   CFU members' perceived level of bushfire preparedness

Forming CFUs: who joins and why?

Perceived risk

The research found that those involved in CFUs recognised the high bushfire risk in their local area. Many had experienced bushfires in the past and were likely to have defended their homes using their own resources, often shared between neighbours. They had discovered that their buckets, mops and garden hoses were insufficient. Those who had not experienced bushfires first-hand were likely to have been informed of the risk by their neighbours or were aware of the fire-prone nature of their surroundings. Many became involved with a CFU due to a strong motivation to protect their own homes and property; obtaining great personal benefit from a minimal sacrifice. Surveys of non-CFU members found that individuals often perceived the risk but did not intend to stay and defend their homes in the event of a bushfire, did not feel they had time to join or considered themselves unfit or too old.

Social setting

Areas with CFUs were also likely to exhibit other forms of community action (or had been in the past), such as neighbourhood watch or bushcare groups, suggesting an existing spirit of community action and co-operation. The high-risk status of cul-de-sacs in the urban/bush fringe appears to coincide with a greater degree of social interaction and shared identity compared to areas along busier roads with less opportunity to meet or interact with neighbours. Many CFU members considered their social bonds to be tighter than in surrounding areas due to their spatial setting, similarities in age, family development, background, greater community stability and shared bushfire experience. This initial bond, accompanied by a higher perceived risk and community spirit, is a likely formula for the formation of CFUs.

How do CFUs improve community safety?

Social capital

All evidence suggests that becoming involved with a CFU allows members to feel greater connection with their immediate neighbours. While the initial core are likely to have been already familiar, expanding the network in order to maintain an active CFU appears to involve a wider range of people with more diverse skills and attributes. Many learn to trust each other's ability, feeling that 'looking after each other' would become increasingly important as they become older: 'being part of a CFU certainly has secondary dividends' (focus group respondent). NSW Fire Brigades have focused on individual streets or parts of streets at particular risk due to poor access, topography or proximity to high fuel loadings. The areas targeted appear to have the local community necessary to actively defend them.

Empowerment

Many focus group and survey respondents stated that prior to the CFU they had felt helpless and unsure as to what they should do in the case of a bushfire. It was commented that before the arrival of CFUs, fire incidents had been disorganised, the fire services had come and gone with little interaction with local residents. Tensions had developed as unfamiliar fire crews gained access to properties and roofs with little regard for their occupants. For some, the trauma associated with previous bushfires had generated a heightened state of anxiety. Some also felt that the declining number of fuel reduction burns in recent years had increased the bushfire risk but their ability to protect themselves had stayed the same.

Since receiving the equipment and training, most respondents had gained confidence in their ability to organise themselves and defend their homes. Although often accepting that it was impossible to be 100% prepared, access to knowledge and resources and integration into fire brigade operations let many feel that they had become a formidable fighting force against the bushfires: 'You can go to war with a slingshot, or you can go with a gun … having a CFU is a bit like that' (focus group respondent). The possession of identity cards[2] was an extra boost as it increased the likelihood that CFU members would be permitted beyond roadblocks and allowed to stay and defend their properties.

Local knowledge

A combination of greater social interaction, on-site training and practical problem-solving has helped enhance local knowledge among CFU members. This ranges from knowing each other's levels of preparedness and resources, e.g. pumps and static water sources, to knowing the best equipment configurations for various situations and the status and whereabouts of other residents. Being able to organise and communicate this information is useful not only for the group but also for any emergency service personnel operating in that area.

Community resilience

It was generally felt that the skills, knowledge and experience gained through association with the CFU are beneficial to the resilience of members. However, respondents frequently identified the firefighting equipment as integral to a feeling of independence and self-reliance. In some cases, respondents felt that they were no longer a burden on the government or emergency services as they were more able to look after themselves: 'We take the pressure off the government because we are not going to stand there weeping because we didn't get support. Now we can do what we need to do, with the equipment we need, with the support we need' (focus group respondent). Some people felt that having a CFU in their local area also made a positive impact upon non-members, as it raised awareness and gave a focus for community efforts both in preparation for the fire season and in actively defending the area during bushfires.

Prepare, stay and defend or leave early policy

Many respondents intended to stay and defend their homes whether or not they were part of a CFU. However, they felt that having training and equipment was an opportunity not to be missed. A minority view was that the CFU and training represented only a small part of bushfire preparedness – fuel reduction, house protection and wider community awareness were all important facets of CFU involvement: 'It's quite easy to think you should [plan] but it just doesn't get done … [being in a CFU] makes you plan' (focus group respondent). In general, being involved with a CFU had boosted individuals' confidence that they could stay and actively defend their homes.

The evidence from CFUs that have defended their homes from fire is positive. The teams work well together and benefit from understanding fire brigade operations and procedures. A more detailed knowledge of pre-fire preparations, fire behaviour, likely ignition points and each other's strengths and assets in a highly localised area all appear to have contributed to the successful defence of homes and property.

What problems have been identified?

Stay and defend or leave early policy

A recurring problem among CFU members relates to the group's overall bushfire preparedness. It appears that attention is sharply focused upon equipment operation, with little in the way of more general preparation: 'I don't know if we've got that great a plan; we would probably work it out on the day' (focus group respondent). Many survey and focus group respondents suggested a 'wait and see' policy in terms of the actions they intended to carry out, with most effort concentrated on extinguishing spot fires around the home. Individual decisions whether to leave a threatening situation were largely unclear; depending upon conditions or left to the advice of the authorities. A worrying 10% of survey respondents stated that their families would 'stay at home as long as possible then evacuate' (coded survey response) in the event of a bushfire in their area.

Many respondents felt that their training had not adequately prepared them for what to expect during a real bushfire, leading to concern at their lack of psychological readiness and how they would cope with stressful situations and dilemmas such as fires affecting multiple properties at once: 'When we get hit it's going to come from the north-west and it's going to come through in about 10 minutes and we won't have a clue ... we'll be in trouble' (focus group respondent).

Community integration and operational logistics

A number of points were raised with regard to the structure and make-up of CFUs and how they developed within the local community. Interest in CFU membership and training is often high after bushfires, but this tends to wane as the perceived risk is reduced. The effect is that CFUs can lose all members or that only a core group remains active. This core tends to share more similarities and, over time, may be regarded by other community members as an impenetrable clique, described by Putnam (2000) as the 'dark side' of social capital. Such barriers to community integration may be exacerbated if fire threatens a neighbourhood, as CFU members may resent the (attempted) involvement of untrained and uncommitted neighbours, while the neighbours may feel alienated and vulnerable.

There was anecdotal evidence of difficulties between CFU groups and neighbours when prioritising properties to be defended. Evidence also suggests that local people are often aware of a CFU's presence but not of their remit, potentially leading to a false sense of security and lack of overall preparedness. If a CFU is pressured to go beyond its designated role during a fire incident and there are dilemmas over whose houses to protect, the potential for conflict is high.

Training and development

A number of respondents expressed concern that the growth of the CFU movement had not been accompanied by adequate increases in support and administration. Some felt that the more personal bottom-up focus had been lost as communications with the NSW Fire Brigades became more difficult and training was provided at large training days, sometimes involving 500 volunteers travelling to a single location: 'Since the changes have come in there is a feeling we have lost the thread of what it's all about' (focus group respondent). Many expressed concern that the kind of training they received at training days was not appropriate to their local situation, minimising the nurturing and transference of local knowledge. The travel was also a disincentive for some volunteers, who no longer attended training days (Ch. 12 this volume has further information on training concerns).

Conclusion

Overall, the CFU movement has created many positive effects at the urban interface, among them a sense of empowerment and community resilience, greater preparedness and awareness and an increased knowledge of bushfires and NSW Fire Brigades/RFS bushfire operations. At this stage of the analysis, three points appear to be significant to the effectiveness and future management of the CFU movement.

Attracting involvement from high-risk groups

It is clear that the NSW Fire Brigade's reliance upon proactive community groups for CFU membership has been responsible for the program's success in sustaining interest and developing highly committed and motivated units. However, a criticism of the CFU program, expressed by individuals involved in other types of community bushfire education programs, relates to equipping groups that may already be well-resourced, aware of the risks and operating well as a community. The extra support provided by NSW Fire Brigades is a positive step in ensuring the organisation and preparedness of certain community groups, but there are important questions about the resource efficiency and social acceptability of supporting them instead of less proactive, aware or articulate communities for whom the risks may be much higher.

It should be considered, however, that the strong community spirit and volunteerism that the CFU formula has tapped into may not be as successful if the NSW Fire Brigades were to approach more dislocated and high-risk communities. Involving such groups and effectively closing gaps on the urban interface would require a far higher input of awareness-raising, time and resources for recruitment and maintenance. Building social capital in areas of low awareness or community integration may go beyond the current remit and capacity of the NSW Fire Brigades, however, the approach may prove more equitable and effective if it can reach the areas most at risk.

Broadening the focus of bushfire preparedness among CFU members

CFU members' heavy reliance upon their equipment may bias their overall preparedness and ultimately reduce levels of resilience. Much of this attitude is born out of experience of being poorly equipped while fighting fires. If they are given hoses, overalls and water pumps, some individuals consider that they have sufficient protection from bushfires. This stance has obvious limitations and dangers, for example, CFU members appear to neglect more detailed action plans for their individual or family circumstances, concentrating instead on their operational role as a unit member. This was evident in the 2006/07 fire season when CFUs in the Blue Mountains deployed hoses, pumps and personnel admirably, but neglected to close windows and doors on individual properties (Munsey, pers. comm. 2006).

The problem of tunnel vision may be exacerbated if members are unaware of or closed to the possibility of co-operation with the wider community or do not create contingencies for a range of situations or developments. Loss of water pressure or electricity or the absence of key individuals from a CFU could seriously compromise operations, making the group extremely vulnerable. Focus group information highlighted the fact that individuals had rarely considered detailed fall-back plans, with many survey and focus group respondents stating that they (and/or their families) would leave if they considered the situation to be too dangerous. This attitude departs quite significantly from the 'prepare, stay and defend or leave early' advice.

Managing the existing membership in a sustainable way

Much of the success of CFUs depends on the management of motivated and proactive community groups. Maintaining this interest is vital if the CFU program is to remain effective and operate as a well co-ordinated wing of the NSW Fire Brigades. However, managing large numbers of volunteers is already straining the limited administrative and management staff. Changes to the running and organisation of the program, while making economic and logistical sense, alienate some groups from what they perceive as the main reasons for their participation. Without the close personal relations and locally focused training that prevailed when membership numbers were lower, it is likely that individuals will begin to question the commitment of the NSW Fire Brigades to the scheme. At present, many people consider that large regional training days and repetitive training drills require a significant effort for little reward. This is felt particularly by groups that have existed for some time. A progressive style of training in a local and familiar setting would be far more beneficial. This is only one example of the kind of issues that may arise as the NSW Fire Brigades attempts to balance the needs and preferences of volunteer groups with safety protocol and scarce resources.

Further research and policy relevance for agencies

- A number of elements of the CFU approach have been highly successful at gaining volunteer interest. Further research should investigate how this can be applied to other voluntary organisations.
- CFU members tend to share social, economic, experiential and educational attributes, a situation that yields committed and well-prepared volunteers. However, methods must be found to engage with and sustain a volunteer relationship with less aware or cohesive groups at the urban interface.
- The popularity of the CFU movement and its growth has raised questions regarding future management and the allocation of resources among high-risk groups at the urban interface. It will be a challenge to maintain effectiveness and ensure the most vulnerable are identified and included.
- The program has been successful at developing a sense of empowerment and community resilience. However, there is a need to widen members' focus from equipment operation to a more holistic preparedness approach.
- The program has not defined its saturation point, i.e. the number of units and volunteers that can be sustained financially and safely in a difficult environment. By 2007 there were over 6000 volunteers, with 9000 volunteers forecast by the end of the decade. Research to determine the number of volunteers that the NSW Fire Brigades are capable of managing with their current administrative and financial constraints would benefit all stakeholders.

Acknowledgments

The research discussed in this chapter was carried with the sole financial support of the Bushfire Co-operative Research Council's Project C6, 'The "prepare, stay and defend or leave early" policy: validation and implementation'. The authors would like to thank NSW Fire Brigades for their assistance in carrying out the research, in particular Commissioner Greg Mullins, Terry Munsey, Karl Deusing and Rachel Scott. The data, views and information could not have been gathered without the help and support of the CFU members who took part in the on-line survey and follow-up focus groups. We would also like to acknowledge Professor John Handmer as Director of Program C and leader of this particular research.

Endnotes

1 As part of the preliminary research, 10 semi-structured interviews were held with individuals involved with managing CFUs at corporate, logistical, administrative and operational levels as well as individuals involved in separate community education and training schemes. Although not used explicitly in this chapter, the interviews were important in defining a background to the issues discussed here (see Lowe & Haynes in prep. for more detail).

2 CFU members have been issued with photographic identification cards since 2005. It is intended that cards will be issued to all members, including those who joined before 2005.

Chapter 4

The concept of local knowledge in rural Australian fire management

Jenny Indian

Chapter summary

The concept of local knowledge in fire management has long been recognised as important. After the devastating 1939 fires Judge Stretton made a key recommendation in the Black Friday Royal Commission that local knowledge be further developed and utilised by forest officers. The concept was again acknowledged as a valuable tool after recent fires (Esplin et al. 2003; COAG 2004; McLeod 2003; Parliament of the Commonwealth of Australia 2003; Government of South Australia 2005), specifically in the COAG Report into the 2003 Victorian fires and a report used by fire agencies and managers. Despite this, the significance of local knowledge in bushfire management, its meaning and practical application remain vague.

This chapter explores the concept in greater detail and seeks to define local knowledge and its role in fire management. This is not simply an academic exercise but one which considers the application of these traditions and experience. The role of local knowledge is discussed not only as a tool in fire planning but as a crucial part of community engagement, allowing rural communities the chance to play a more active role in fire management. The impact of this involvement and acknowledgment is considered in relation to the resilience of specific rural communities.

However, accessing and using local knowledge is not without difficulties: who holds this knowledge, where is it to be found, is it correct, how can it be verified? Oversimplification of the term, its use and understanding are all potential dangers. The historical relevance of the knowledge, the user's perspective and the place of local knowledge in the broader issues of fire strategies may complicate inclusion of this concept in bushfire response, but it remains fundamental to both future fire management and genuine community engagement.

Introduction

The concept of local knowledge is widely embraced and recognised as important in the debate surrounding fire management (post-fire public inquiries Esplin et al. 2003; COAG 2004; McLeod 2003, Parliament of the Commonwealth of Australia 2003, Government of South Australia 2005) particularly after the extensive 2003 fires throughout the alpine country of south-east Australia. During January of that year dry storms ignited a number of fires in the north-east of Victoria and Gippsland which, with deteriorating fire weather, developed until the firefronts joined. The fires burnt 1.12 million hectares of land, including over 108 000 hectares of private land. No lives were lost as a direct result of the fires (Esplin et al. 2003). In considering the resilience of these high country communities to fire and in speaking to individuals with direct experience, discussion often returned to the question of local knowledge, its use, dismissal or simply what it involved. It rapidly became apparent that there was much more to be learnt about this subject, and that further research into its role in future fire management was called for.

Although the term 'local knowledge' is often used, interviews, discussions and participant observations have shown that the concept remains potentially misunderstood. Though initially seeming quite simple, local knowledge can present as a complex mixture of observations, thoughts and reasoning based on local experience and tradition. By its very nature it is a loose term – human beings are themselves complex, their communities notoriously so and their communication, understanding and knowledge far-reaching. For the purposes of this chapter, local knowledge is considered as information based on tradition, personal observation and experience of a particular geographic location. How this information is held and learnt within a community, both first-hand and passed on, will also be a research focus. All knowledge has a context and, as such, who the expert is depends on the circumstance (Yli-Pelkonen & Kohl 2005). We can delve beyond that and suggest that local knowledge involves a degree of understanding over and above simply knowledge. Information exists and is received, but interpretations vary. Indeed, it has been noted that knowledge is not something an individual has 'more' or 'less' of; rather it reflects the specific forms of practice undertaken in daily life. Thick in some areas and thin in others, knowledge is embedded in daily political and environmental activity (Robbins 2004). Those with genuine local knowledge are able to offer thoughts and advice grounded in experience and tradition and therefore underpinned by understanding. The tacit, almost elusive, nature of local knowledge must also be recognised as contributing to the inherent difficulty of isolating the concept.

This chapter suggests that local knowledge has the potential to play a crucial role in future fire management within Australia and beyond. However, it is important to recognise that the function of local knowledge will alter according to the specific phase of fire management and, as such, the role of this tool will therefore vary. Within certain phases of fire management local knowledge may play a fundamental role, for example in early planning and aspects of operations. Within more strategic operational response situations, the function and value of local knowledge may be more difficult to isolate and identify. Importantly, it is felt that the successful transfer and use of local knowledge has a significant role in improving community relations and trust, particularly community participation, and in fact underpins many aspects of successful and genuine community engagement.

Research in this area continues and extensive interviews have been conducted with individuals, local government officers and government and non-government agencies. Focus groups have brought together a range of locals and newcomers within these communities, with a diversity of thought and opinion. Participant observations have been conducted at community meetings and current and more established literature has been reviewed extensively. Dialogue with others involved in this field of research overseas is also ongoing.

The concept of local knowledge is explored in greater depth as its inherent complexity deserves careful and critical scrutiny. It is hoped to raise the profile of local knowledge – particularly in relation to the 2003 fires in the alpine country in Victoria, New South Wales and the Australian Capital Territory but also well beyond – and to extend current thinking and discussion of the concept. This chapter is written for fire agencies and managers and directed at those with the capacity to explore and understand the role of local knowledge and its application in future fire management.

The idea of local knowledge

Traditional ecological knowledge (TEK) is a form of local knowledge as ancient as the hunter-gatherers, yet the term only came into widespread use during the 1980s (Berkes 1993). This form of knowledge represents experience acquired over thousands of years of direct human contact with the environment. However ancient it may be, TEK is difficult to define and can be viewed as ambiguous – societies change and new practices and technologies develop, making it difficult to specify how much and what kind of change would affect the labelling of a practice as traditional (Berkes 1993). TEK can be a vast accumulation of knowledge and understanding. In the alpine areas of Australia, for example, large fires were apparently extremely rare before European occupation; the movement of Europeans into the area brought a massive increase in both the frequency and intensity of fires. Studies indicate that fire frequency and intensity in the Alps under Aboriginal management were far lower than in the rest of south-eastern Australia (Zylstra 2006). The landscape of this area is therefore irrevocably altered –perhaps compounded by climate change – making any attempted re-creation of the original ecology virtually impossible. This is not intended to reduce the importance of Aboriginal TEK in the area of fire management and understanding; indeed, that knowledge should be sought and considered as part of the community engagement process.

Local knowledge by definition can be intensely specific, applicable within a very small geographic area and so potentially limited and problematic in application. However, its successful use and transfer within a district may have generic application elsewhere. How local knowledge can be harnessed and practically applied and why this is important to the process of community engagement is at the core of a genuine understanding of this concept. Local knowledge can be riddled with subjectivity, coloured by self-interest, value-laden emotions and potential weaknesses. It can be used to air long-held grudges, general distrust of authorities and personal gripes. While remaining aware of this, it is worth noting that being supposedly value-free does not necessarily mean being totally objective. In dealing with people and their long understanding and experience, we may stray into the realm of intuitive knowledge, itself an area of extensive research and controversy. Local knowledge is difficult to measure and test quantitatively as it involves 'local values; anecdotal, observational experience; colloquial terminology; the all-but-invisible background of relationships, behaviours and kinship structures that shape people-environment relations' (Howitt 2003: 3). Thus, the research follows a more constructionist philosophy where the social phenomena investigated may not be directly observable and perhaps only implicit and approximated (Robson 2002). This approach favours a more sensitive anthropologically based qualitative research design involving unstructured in-depth interviews, focus groups and participant observation.

However, local knowledge need not be set up against scientific or expert knowledge. it can be viewed as complementary (Mahiri 1998). For example, where scientific knowledge can dictate overarching policies and practices, local knowledge can guide local, more practical applications. Herein lies the dilemma of local knowledge, its apparent weakness and inherent strength. Local knowledge underpins the knowledge interface between experts and locals and thus plays a pivotal role in communication by promoting trust and co-operation. What is absolutely fundamental is the acceptance and understanding that local knowledge must be subject to the same scrutiny as all knowledge collected. Simply because it is deemed local knowledge does not mean that it is necessarily more suitable than outside advice – it must be fit for the purpose.

In considering this area of research it is tempting to indulge in lost rural traditions, glorify the past – the 'good old days' – and lament the demise of self-reliance in changing rural communities. All too often 'Collective memory simplifies; sees events from a single committed perspective; is impatient with ambiguities of any kind; reduces events to mythic archetypes' (Bean cited in Manne 2007: 26). The changing demographic of rural populations, the questioning of the sustainability of farming practices and perceived overarching bureaucracy and regulation can be viewed as disruptive elements to the apparent rural idyll. However – particularly in relation to fire management – many aspects of improved scientific knowledge and technology are enormously positive, and often enable the continued existence of isolated rural communities which may otherwise decline into oblivion. Greatly improved communication, understanding of fire behaviour and constantly updated equipment must be acknowledged as crucial to fire management. Local knowledge should not be seen as an opponent to current fire management practices, but as another tool for use by fire managers and agencies, based more prosaically on human understanding and history (and therefore perhaps misunderstood and underused) and yet fundamental.

Where to find it; what forms it may take

Rural communities

Communities throughout the world are notoriously heterogeneous and rural settlements are no exception. Within many rural towns and regions are those with experience and knowledge of their district, and how it interacts both socially and physically with fire. For example, knowledge of the surrounding landscape, the problems it may present, safe access points or areas of danger; knowledge of the local community, who lives where, their age, health and vulnerability; the fire history of that particular district and the greater region; key members of the community with specific knowledge and understanding; weather and landform patterns; what to expect of the local community and, importantly, how to approach the informal leaders within it. This knowledge is simply part of the community life, and in isolation may remain so. However, it can be enormously helpful to those seeking to know more about the area.

Government departments and fire agencies

It is widely acknowledged that the local knowledge of firefighting personnel during the 2003 fires across several states was under utilised (Esplin et al. 2003; McLeod 2003). Country Fire Authority (CFA) volunteers come from within the community and many have wide experience of their specific locality. In addition, the role of local knowledge within various government departments – for example, the Department of Sustainability and Environment and the state Parks Departments – has long been valued within rural communities. The value of continuity of employment and location is again beginning to be acknowledged.

Harnessing

Harnessing local knowledge from within a complex and often disparate rural community presents obvious difficulties. Local knowledge can indeed be very local and as such not transferable and there is no template for the accruing of this tool. Talking to locals in their own environment is an obvious beginning, as is providing circumstances where people can feel comfortable in coming to you: 'how citizens are invited to participate in disaster management is critical to the success of that participation' (Pearce 2003: 218).

Using other relevant and successful models from rural Australia may go some way to alleviating the situation and help to progress research in this area. Landcare (a partnership between community, farmers, government and business working toward the practical protection and repair of the Australian environment) is one such model which could be explored and developed to provide further ideas for the potential resourcing and support of accessing and using local knowledge. Nationally this has proven a very successful approach both in terms of outputs and in galvanising local communities. This approach does not impose project models or outcomes on the community, but allows participants to create projects and goals suited to their local community and landscape. It must be noted, however, that fatigue is currently very apparent within Landcare, with comments about 'drowning in paperwork' and the increasing complexity of grant applications taking too much time (Curtis & Cooke 2006).

It is important to recognise that many rural Australian communities are struggling – the pressures of climate change are affecting them dramatically, and the cost of living in more isolated parts of the country is increasing while incomes decrease markedly. To tap into the wealth of community knowledge and to utilise this information efficiently and well, the process should be driven from within the community rather than imposed from above. This must be handled carefully, with discretion and respect. It will take considerable time and effort as endless voluntary work is beginning to take a heavy toll on rural communities (Graeber 2006). However, it may be necessary to make the additional demands of meetings and paperwork on an already stretched rural community while tapping into and utilising local knowledge. It must be undertaken with care and precision. As with any government or agency initiative, avoiding unfamiliar language and a totally rigid or prescriptive process are crucial to success.

Figure 4.1 Liaising with community members, on private land in the Snowy Plains, NSW, March 2006. © Bushfire CRC.

Key players

Influential and respected people exist in nearly every community. These key players can provide an enormous amount of information, contacts and facilitate networking throughout the community. In several small isolated towns in the high country of Victoria and NSW, during the 2003 emergency, the initial firefighting effort was virtually run by one person, with considerable support from the community. However, we must be careful to ensure that all stakeholders are included, not just the most vocal. All knowledge sought and received must be considered and the danger of the least economically powerful participants being (further) marginalised must be recognised. Robbins (2004: 18) suggests that 'struggles that either seek to criticise or vindicate local knowledge ... are less useful to understanding socio-natural processes than those that embrace the positioned and partial nature of all knowledge'. The most useful local knowledge may come from the least vocal – the skill is in discovering just where this knowledge is, gaining the trust and confidence required to obtain it and sifting through the information to ensure its application is most effective for all. This may not be easy – those who challenge the status quo may be the very ones who have much to offer and who can provide new and different perspectives but they themselves may present challenges. They may present as difficult and cynical but provide a completely different perspective with their particular knowledge and experience of the area.

Existing community groups

These groups may be apparently unrelated to the specific knowledge sought, for example local recreation reserve committees, cemetery trusts or school councils, but they often involve a number of key players from across the region and thereby allow broad access to a number of local people. Many of these groups are very pleased to include a guest speaker on their agenda and contacts can be made for future reference.

Resourcing

Adequate resourcing is all-important and the use of local knowledge is a resource-intensive exercise. While this must be largely driven from within a community rather than imposed from above, support must be provided. Additional and specific staff from agencies who are charged with accessing and documenting aspects of local history, traditions, social strengths and weaknesses is one obvious approach. For example, in Victoria, the Department of Sustainability and Environment has established regional officers whose primary function is to liaise with locals with a view to improved community engagement.

Applications

The use of appropriately derived local knowledge can only benefit future fire management, whether this is the explicit transfer and use of otherwise unknown or misinterpreted local geographical knowledge or through general fostering of trust and co-operation between community and agencies. In times of emergency it is crucial that local confidence in fire agencies and management is not compromised, as this can quickly deteriorate into anxiety, uncertainty and often mistrust. If that begins, long-term alienation and disquiet can occur. There are innovative opportunities to harness and utilise local knowledge at regional and international levels for both fire brigades and government departments. They include the following.

- The training of specific fire units – those from completely different geographic terrain could be given training in districts very different from their own. Those from very isolated rural districts could undertake some training in the rural/urban interface, thus tackling some of the increasingly common problems with this area. Individuals can broaden their experience and be taken out of their comfort zones, allowing teams to feel more confident if their brigade attends fires outside their region, and indeed in another country. That often happens in Australia with the use of US, Canadian and New Zealand fire personnel. The opportunity exists for outside teams to rub shoulders with the local knowledge of another district and absorb first-hand information about that region which may prove useful.

• Include local brigade personnel in the decision-making process. Instances occurred during the 2003 emergency in the Victorian Alps where local brigade personnel felt locked out of decision-making by visiting Incident Control Teams, without adequate reason or explanation. This causes immediate (and long-term) ill-feeling – the ripple effect through brigades and local communities can be enormous.

• At certain times, however, the displacement of local brigades may be the only way to respond in an emergency situation. It is then particularly important to ensure that the displaced nevertheless feel that they and their facilities are respected.

Table 4.1 illustrates this approach more simply.

Table 4.1 Strengths and weaknesses of local knowledge

Stages of fire management	Strengths of using local knowledge	Weaknesses of using local knowledge
Listening	Early communication and contact with locals Community involvement indicates a long-term investment Broadening of information base Promotes trust and co-operation	Hard to access and verify May not be altruistic Too local/narrow for use May encounter conflict from and within community Resource-intensive
Planning	Potential for new perspectives Building community involvement and confidence Greater confidence as decisions made based on verified/checked local knowledge Utilising existing information, not reinventing the wheel Allows early organisation of specifics, e.g. to ensure presence of a local in IMTs May aid the application of broad government/ agency policy	Limited in application; impossible on broad scale May be gaps in knowledge interface (between locals and agencies) Some people and views may be excluded
Operational response	Taking local community with you; greater understanding of decisions made due to early inclusion Increased communication and involvement results in better fire awareness and understanding Potentially greater confidence, e.g. when using maps checked by locals Strengthening community involvement and understanding of emergency also increases resolve and underpins resilience	Use of local knowledge less tangible/apparent causing some lack of trust to remain

Though initially very resource-intensive, the accumulation and use of local knowledge is a long-term investment which has the potential to aid all phases of fire management, in particular the final operational phase when the harnessed information is used to reinforce community participation and engagement.

Examples from the 2003 Alpine fires

Specific examples of the use of local knowledge from the 2003 fires in the Australian high country help to show how observation and use of tradition and experience can assist in times of emergency. Obviously, a more measured approach is tapping into and documenting this information prior to an event and applying it as part of an overall planned approach during the emergency (see Table 4.1). More common examples include the following.

- Locals assisting with mapping, for example location of access tracks and problems with terrain. This is particularly effective prior to any emergency to allow landholder participation, thereby reducing the potential for poor frantic decisions and the resultant negative backlash.
- Older brigade members briefing firefighters both in the shed prior to departing and actually going on the trucks to provide on-the-spot guidance with access, terrain and potential dangers and, importantly, to negotiate with local landholders.
- A list of potentially more vulnerable community members, for example older and frailer people, those with young children, those less physically capable or simply living alone and potentially unable to cope. Once established, this document can be used to check on and warn residents of the approaching fire. Care must be taken to ensure privacy concerns are not breached.

A more negative experience from 2003 provides a clear example of how a lack of local knowledge can very quickly undermine the trust of a community. In an interview, a local government officer described how employees were advised to go home to protect property. However, a police roadblock prevented people from returning home. The police officer in charge had just arrived from a major city, hadn't been adequately briefed, was unsure of the exact direction of the fire and could not suggest where those redirected from the roadblock go for safety as he didn't know the area at all. The local government officer returned to the office, copied several maps and took them to the police officer and his counterpart at the other end of the roadblock. Leaning on the bonnet of their vehicles, she enlightened them as to location, places of safety, fire direction and other very relevant local information. In doing this she also managed to get home safely to protect her own property. This situation undermined local confidence and had the potential to seriously destabilise the community, in particular during the height of the emergency. However, the situation was reversed through the provision of appropriate local information.

Pitfalls and potential dangers

Considerably more work is required to ensure that the future use of local knowledge in fire management is most effective. However, there are several potential dangers, some of which are very obvious.

Reaching community consensus

Many varied views will be encountered but not everyone wants to be involved and contribute. Who in the community holds information? Is it reliable or is it dated, based on opinion rather than fact, charged with emotion and sentiment? Is it totally subjective rather than objective and is this necessarily a problem or does local knowledge, by its very nature, involve degrees of subjectivity? What is perceived as local knowledge within a community must be examined in the light of genuine altruism – will the information help the community for the common good, or is it provided by an individual or small group to help an individual, possibly even at the expense of others? Is it coloured by self-interest? Is it too narrow and, if followed, will it actually jeopardise the broader fire effort? Does the information actually provide a complete picture? Judge Stretton (1939: 7) noted that 'The truth was hard to find. Accordingly, your Commissioner sometimes sought it (as he was entitled to do) in places other than the witness box. Much of the evidence was coloured by self-interest. Much of it was quite false. Little of it was wholly truthful'.

Who is to judge what the balanced view is and what is in the best interests of the wider community? How can this compromise be achieved? Outcomes from seeking local knowledge are not necessarily predictable or tangible; verification may be very difficult and take time and resources. If 20 locals are invited to attend a meeting, 20 different views may arise. This is not a simple tool. In considering local knowledge we must always proceed with caution.

Use of language

Rural observations and use of language can vary from those used by people from a more urban background. Overuse of acronyms and scientific terms can be confusing and alienating. To work effectively together, language must be understood by all and respect be shown for variations. Less articulate people are still worth listening to and their views must be sought with suitable methodologies. Poor choice and use of language early in the process of harnessing local knowledge can be alienating, causing potential participants to shut down and dismiss the process as a waste of time.

Exclusivity

Local knowledge can bring a position of power in a community – some have it and some don't. Not only does this have the potential to further fracture the community but it can lead to ego dominating common sense. People who are perhaps unsure of the worth of their knowledge may remain silent and, most importantly, not engage in a process which could otherwise be a learning experience for all.

Inherent parochialism

All too often in situations such as fire management, 'sides' develop and become polarised and the resulting conflict affects everyone. Ironically, both 'sides' often have much in common – a genuine sense of caring for country – yet inherent and entrenched prejudices won't allow that recognition. Forms of 'simple science' can be used by locals, particularly if they are already disgruntled, for example 'this is what happened here last time and therefore we can extend that further, to cover the entire region'. Land management issues often fall into this category, with local communities disputing extensive scientific data. This will further complicate any meaningful dialogue and, when questioned, can alienate. Prejudices can become entrenched, dominating and distorting dialogue.

Misinterpretation

The term 'local knowledge' and its role can be confused and misused as something of an elixir for all the ills befalling communities before, during and after a major emergency. Not only is this misleading and confusing, but it can deepen any rifts between communities and government agencies, causing blame and a negative backlash which further complicates recovery. To be truly effective, the role of local knowledge needs to be explained and understood clearly.

Local knowledge is only one tool available to fire managers and must be viewed as such – it needs to be scrutinised as any information should be when used to inform decision-making. It cannot therefore necessarily inform management practices as part of a blanket approach and should not drive these decision-making processes. It must be viewed as situational and contextual – specific, applicable at certain times, in certain locations and in particular ways. It may not be geographically transferable.

Aspects of local knowledge cannot be preferred over expert or scientifically based information purely for the sake of public popularity. The enormous benefits of choosing to use a combination of approaches, of incorporating all possible tools in fire management, must be viewed in the light of potential misuse and dangers. Local knowledge should be used to feed into and complement scientific study rather than be an 'either/or' choice. It should be viewed as a form of hybrid knowledge – one form of knowledge should not be privileged over another.

... and community engagement

Harnessing local knowledge for practical application means that we are seen to be trying – it allows people to feel valued and part of the decision-making process. This builds confidence within the community and allows the growth of trust in Government agencies; the way to have an informed public opinion is to inform and involve the public. Though these outcomes may be less tangible they may in time provide the greatest benefits – improving community relations and resilience, strengthening community resolve and building a better knowledge and understanding of fire in Australia.

Two-way listening is fundamental to this process; there must be a genuine shift away from community participation being just another vehicle for community education. All of this takes time but allows a broader perspective which will contribute to better long-term decision-making. By actively harnessing hindsight we are building on the lessons learnt from emergency situations.

It is clear that the most disempowering aspect for those involved in events such as the 2003 fires is complete dismissal of their input – the apparent lack of consideration and the acknowledgment that their knowledge and understanding is considered irrelevant. Obviously, during the height of the fires time is of the essence and decisions are made rapidly – this is not the best time to begin working with the community and tapping into aspects of local history. Local knowledge is not a spontaneous tool and cannot be used in an ad hoc fashion. It must be part of a long-term process, gathered over time and fostered within a community. Rural communities are complex and vary enormously and, while they share common themes and aspects, a 'one size fits all' approach will not work. Each region, district and community has its own history and sense of country. This must be acknowledged and respected and the specific attributes of that knowledge and understanding be openly recognised and valued at all levels of fire management.

Conclusion

In exploring the concept of local knowledge it is apparent that its use in fire management is both complex and controversial. As with many aspects of community involvement and participation, it must be scrutinised carefully and, though initially very resource-intensive, the accumulation and use of local knowledge must be acknowledged as a long-term investment which can aid all phases of fire management. It is crucial that the term and all it represents not be considered as a magic bullet which can overcome all perceived problems in fire management and rural communities.

By taking the time, providing the resources and being involved in the use of local knowledge, fire managers and agencies are investing wisely, acknowledging the wealth of available experience and developing the skills necessary to ensure the reliability and effective application of this tool. No longer can the experience and tradition of rural communities be held at arm's length – all levels of fire managers and agencies need to embrace this understanding, rub shoulders with those who have it and use it willingly and innovatively in their planning.

Chapter 5

Social contexts of responses to bushfire threat
A case study of the Wangary fire[1]

Helen Goodman and Mae Proudley

Chapter summary

Using the Wangary fire in South Australia (January 2005) as a case study, we aim to deepen our understanding of how families respond to a bushfire event. This chapter focuses on gender and relationships including the following specific issues:

- What influences the decisions and actions of families when threatened by bushfire?
- Was the 'prepare, stay and defend or leave early' policy a feature in the accounts householders gave of their decision-making on the day?[2]
- How does the presence of children influence preparedness and response to bushfire?

Semi-structured interviews were conducted six months after the Wangary fire by three researchers from the Bushfire Co-operative Research Centre at 17 households across the fire-affected region.

This chapter proposes that concepts of role and identity within the broader framework of 'community' can be useful lenses through which to examine issues of community safety in relation to the threat of bushfire. Community safety messages, while often aimed at individuals, need to take into account social interactions at the household and neighbourhood level. Some households managed to remain more cohesive in their fire response, thus reducing their exposure to the threat of the fire.

This research supports the expansion of initiatives which enhance community members' capacity to come together in groups to develop and expand their responses to the threat of fire. Volunteer firefighters, who are also farmers, are looked to by some community members for guidance. In the main, they did not voice the language of the 'prepare, stay and defend or leave early' policy which suggests scope for more discussion to understand the reason for this and to explore how it might be altered. In addition, the desire to remove children from perceived danger can be overwhelming and may result in high-risk behaviour.

Introduction

This chapter reports some of the findings from our analysis of 17 household interviews (35 participants) carried out in July 2005, six months after the Wangary fire.[3] Three interrelated issues are discussed: different fire experiences among residents, the relative absence of the 'home as refuge' idea and the influence of children's presence when thinking about preparedness and response. Implications from these three themes are presented for their contribution to dialogue within and between end-users, researchers, community services and community members.[4]

Issues of role and identity are informed by the contexts in which we live, and by our own histories. They contribute to our sense of social belonging and affect the choices we make. The chapter seeks to deepen an understanding of the complexities of preparedness and response to fire in a particular geographic locality and social context. In highlighting some results of exploratory social research of the lived experience and the reflections of 35 residents who faced the threat of fire, we note our respect for those who shared their experiences and reflections. We hope to gain insight into their decision-making in the face of extreme pressure, and to pass on some of that learning to others.

The fire event is known as the Wangary fire. Wangary is on the Lower Eyre Peninsula of South Australia. Its main town, Port Lincoln, has a population of approximately 14 000 people and is 676 km west of Adelaide – it is described as 'remote' by the Australian Bureau of Statistics. The surrounding landscape is largely agricultural, supporting a substantial number of grain businesses and grazing. It also includes a considerable amount of native scrub and forest. The coastal areas support a large fishing and tourist population, and there are several smaller townships inland.

The fire began on Monday 10 January 2005 and was declared contained that evening. On Tuesday morning, ('Black Tuesday') the fire broke containment lines, and claimed nine lives. Two of the nine were men on a private fire unit; seven were women and children, six of whom were fleeing the fire in cars. Ninety-three homes were destroyed, along with numerous other buildings, vehicles and equipment. Over 46 000 livestock were destroyed (Smith 2005). The Wangary fire was a severe fire event, with the grassland fire danger index peaking at over 350 (Gould 2005).

Methodology

A survey carried out by Rhodes for the coroner (Rhodes 2005) gathered 288 responses from residents of the fire-affected area. Two hundred and sixty-two residents were asked if they were willing to speak to Bushfire CRC researchers, 191 agreed. The aim was to interview residents, of diverse age groups, from across the fire-affected region. Two months after the telephone survey, households were contacted and appointments made to conduct the interviews, resulting in a sample of 17 households. The interviews were semi-structured and all began with a request for a summary of what they had experienced on 11 January 2005. Ten of the interviews were with couples (husband/wife), two were with individuals (women with absent husbands) and the remaining five were with families (where adult children/brother were present). Three interviewers were present at 15 of the interviews (two females and a male) while two interviews were conducted by the female interviewers.

Table 5.1 Key descriptors of the 17 households interviewed

	No of households	Renting (R) or Owning (O)	Primary Income: farming (F), retired (R), other (O)	Households with school-age or younger children
Under 30	2	R (1), O (1)	O (2)	1
30–39	1	O	O (1)	1
40–49	3	O	F (1), O(2)	2
50–59	8	O	F(5), R(1), O(2)	2
60–69	2	O	R(2)	
70+	1	O	F(1)	
Total	*17*	*17*	*17*	6

Household characteristics

Of the 17 households, 15 were located within the fire ground. Two were in nearby towns but had farming/business interests, and were on those farms at the time of the fire.

Table 5.1 summarises some of the key descriptors of this group of 17 households. Members of two households were under 30 and one was over 70, with the remainder spread between 40 and 60. Five households had school-aged children or younger, but none of the children were home when the fire-front arrived. Members of one household were renting their property; the others all owned or were buying their property. Eight households derived their primary income from farming, and nine did not. Among the nine were those with off-farm income and/or retirement income.

One theme which recurred through the interviews was the variation between the households in relation to experience of bushfire. The main bushfire experience in this context was also linked to household members being current or past members of the fire service.

Table 5.2 Current or past fire service experience

Association between residence and land	Current or past fire service experience	No fire service experience	Total
Full-time farming (owners)	7	0	7
Hobby farming (owners)	4	5	9
House rental on farm (non-owner)	0	1	1
Total	*11*	*6*	*17*

Of the 11 households with fire experience, it was typically the male who was (or had been) a member of the Country Fire Service (CFS), the South Australian rural fire service. Five of the 17 households had members who had been part of a brigade response to the fire on the Monday evening (the day before the fire), and three included family members among the brigade's response the following day (Tuesday 11 January). Another two were active on farm firefighting units on properties other than their own on the Tuesday. Some interviewees spoke of tension between their brigade role and their community role, such as conflict between local and centralist decision-making in relation to the operational response to bushfire. Some referred to aspects of the operational response to this and a previous fire in the region, referred to as the 'Tulka fire' in 2001. This tension between roles was distressing to several people we interviewed.

In terms of losses experienced in the fire, no families interviewed had immediate members closely related to those who died in the fire. However, some knew the deceased personally, as neighbours, as co-workers, as parents and children of the same kindergarten or school. Some had been in contact with relatives of the deceased on the Tuesday, in their search for missing family members. Two households we interviewed had members who were in direct contact with the two firefighters who perished. Other losses included a home, fencing, sheds, personal belongings, tools of trade, wildlife, community symbols (e.g. Wanilla RSL Hall and a restored historic homestead), animals (mostly sheep), and deterioration in health and mental health. Some reported that the disaster had strengthened relationships and individual self-esteem.

Orienting values and concepts

Materials designed to assist people to prepare for a bushfire event often focus on tasks – the myriad of activities people can engage in, prior to and on the day of a fire, to mitigate the fire's impact. Not so readily available to us was the concept of role.[5] For some individuals, familiar patterns were partly informed by their past fire experience. As one of the female respondents said: 'the women get to do the food, to feed the fellas, you keep abreast of where they are, and the fellas go and fight it and, well, they come home in 24 hours and it's all out … but it wasn't the case for this one'.[6]

In some instances, the assumption of the (usually) male role as director of household actions in responding to the fire was clear and probably life-saving: 'command me what to do', one woman said to her husband, recognising she felt ill-equipped and there was no time for conversation. Other role relationships were more fluid, and men left the choice of what to do to the woman. A few men adopted roles they later regretted, either suggesting a certain course of action or not being more authoritative about the action that was taken.

One of the dilemmas in presenting research findings in this field, particularly where we attempt to understand action from a role perspective, is that role behaviour at the household and neighbourhood level is located at the border of private and public life. How individual households work on a day-to-day basis, or in a crisis, is very much part of the private domain. Yet the way households, neighbourhoods and services respond to the threat of fire is also linked to public consequences. This ushers in a raft of public interventions, seeking to determine different sets of accountabilities. This is most dramatically symbolised in the case of the Wangary fire, where the coronial inquiry (at the time of writing, February 2007) has just passed its 12th month of taking evidence. In this forum, some aspects of what might have been deliberated more privately in past years have become very public.

Role and identity are powerful ideas and result from conscious and unconscious states of mind as well as past and present historical, economic, cultural and social forces. What elements of the social and organisational systems we live and work in enable or constrain the enactment of work and community roles? What are the community safety implications when these enactments are enabled or constrained in these ways?[7]

Responses to the direct threat of fire

Three key themes struck us as we worked to understand some of the issues presented during interviews. These themes are:

- the influence of participants' experience with fire
- the relative lack of discussing the 'home as a refuge' idea
- the presence of children as mediating factors in decision-making.

We discuss these three issues below.

Differential experience with fire

Relationship between experience, risk perception and preparedness

The high rate of association with the Country Fire Service as current or past volunteers among our 17 households meant that there were a number of households with fire experience. This group demonstrated an increased capacity to notice the visible cues of the fire and immediately tap into informal networks to obtain further information. They were more attuned to the weather pattern and its likely effect on the fire's progress.

> We've been planning for fire ban days, and bad weather days, and very bad fire days, but we haven't been planning for *extreme* days … I could see the weather was getting extreme, and I could see it was going to be a horrific day, not just a bad day … My own experience tells me, and I've been fighting fires all my life … I could tell this was something we were really going to be struggling with.

One man who, in our assessment, was the most thorough in his preparation both in the preceding months and on the day of the fire, was not a farmer but had experienced years of responsibility for others in a high fire danger area.

> [Fire] puts the fear of God in you. It never leaves you … Once you've been touched by it, you are more receptive. You pick up on things. I have a pretty healthy respect for what fire can do and how quickly it can happen.

By contrast, one young woman who told us she was 'blissfully unaware' of any danger on the Tuesday morning: 'I knew there was a fire but having never been through a bushfire, it didn't really bother me. I figured – It would be OK. We'll be safe.' She went up Winter's Hill and took photographs before her husband alerted her to the imminent danger to the farm where they had business interests, then together they drove through the fire-affected area to get to the property.

Capacity for allocating roles between household members

On the day of the fire, in some households the male was able to take the role of leader. Where other household members were willing to follow, there was a sense of a cohesive response. Women gave examples of being issued instructions by their husbands which enhanced the overall safety of the household.

> Though it was really something I'd never experienced before and with the trees all alight and everything but we were just so busy and I just had confidence in Bill really, so … that's all there was to it, you just had to get on and do what had to be done.[8]

Sometimes the cohesion seemed to extend the household's ability to assist non-household members, such as one partner leaving directions for the other while quickly checking on an elderly neighbour. Sometimes the serendipitous arrival of another person into this cohesive system further expanded what could be achieved. Our discussions included several instances of this. In one case a neighbour, who couldn't return to his own property due to the passage of the fire, assisted an elderly woman to protect her property. While she drove the tractor he extinguished spot fires with a fire unit on the back of the trailer.

Men sometimes explicitly valued the presence of their partner in defending the property. One man reflected on how he drew comfort from the knowledge that his wife was indoors; he would run in, take a break from the smoke and run out again. He was glad she was aware of his movements, just in case he passed out. Sometimes women ascribed themselves some value in this more intuitive and relational role.

> Every now and then if somebody went out of sight, I was … I reckon I took the role of making sure … nobody succumbed to smoke when we weren't looking sort of thing.

One woman was instructed by her husband to take charge of the tenants in a rental home on their property. This role was difficult as the male tenant had a different opinion on what action to take. To her relief, the tenant complied.

Sometimes an instruction separated household members, such as when women were advised to leave the home and go into a neighbouring town. One woman who had been given this instruction told us she 'felt I would be in the way. I didn't argue with him. He's got enough to worry about'. Other rationalisations women reported were that this decision (being advised to leave) meant that they could keep company with another anxious family member in town. Another reason offered [by a departing woman] was that in taking a car away from their property the vehicle would be saved even if the property and belongings burnt. In a few cases both parties (husband and wife) seemed to convey a sense of ambivalence about their actions and rationales. In one case, nobody questioned the male view that the best advice was to leave.

One young woman told us of her unsuccessful efforts to carry out her own inclination, which was to keep family members together. One member decided he would go and move sheep, while another member in the same household tried (unsuccessfully) to fill the fire unit. The woman said:

> We didn't really have time to communicate … my concern right from the
> start is I don't think it's a good idea to separate because that's how accidents
> I felt happen … so I wanted to stay together … Barry obviously left and then
> Tom went down there to get the fire unit and I was just here by myself, so I
> got in the car and said … look it's not worth it.

In her journey through the smoke and flames she narrowly escaped losing her life.

Fire experience and fire resources

Greater experience with fire generally meant a greater likelihood that households would have the resources to assist in home defence, such as fire units, appropriate protective dress, hoses which withstood ember attack, hose fittings which allowed rapid tank-filling, diesel or petrol pumps for when the electricity went off, and masks to assist breathing. Grain-growers use fire as a crop management tool, and often graze animals to reduce fuel. Fire is also a consequence of farming, with a high rate of fires from the use of harvesting machinery. One farmer summed up the interactive nature of land use and fire: 'We assume it [fire] as an ever present threat'.

Fire experience and expectation of outside help

Those with more fire experience also had a more realistic approach to the likelihood that they would receive assistance from the fire service.

> I said to Mary at some stage, look we can't expect any help because the fire
> has travelled 25 km or so at this point, and there are so many farmhouses
> between here and there that the resources would be stretched to the limit
> and we couldn't expect any help.

This statement contrasts sharply with that by a younger woman with no fire experience:

> I've had the wrong idea … particularly being out here, I've never ever worried
> because I've always just thought if there's fire, there's a fire brigade, I never
> ever thought I'd ever face a fire on my own.

Socio-economics

However, viewing experience with fire without considering other factors would provide an unbalanced account. Socio-economics play a role in bushfire preparedness. Our data showed a relationship between skills and knowledge derived from agricultural work, but we also observed some households where the barrier to response capacity had socio-economic elements (precluding the acquisition of resources).

Gender[9]

We also caught a glimpse in the group of 17 households that sometimes it is the women who are more regularly mindful of the threat of fire but, for various reasons, that awareness does not receive attention from their partners. Two women in our 17 households, whose partners' work took them away from home for considerable periods each year, told us that they had lived with a chronic state of anxiety about fires for many years. One woman said her family referred to her as paranoid, and conveyed the impression that they belittled her concern. Another woman said she was always checking with her husband as to whether the fire unit was full and ready. One of the two households had sold their property between the fire and our interview, and the other was talking seriously about selling.

Box 5.1: Fire experience variability in the community – implications

Messages in relation to community safety often seem to be aimed at the individual, as if they act in isolation from others. It appears that we need to understand more about the various roles taken up in households in relation to community safety issues.

- If we did develop more understanding of these issues, what might this mean for community education approaches?

Some people could use their experience as farmers, hobby farmers and members of the fire service to assist others to prepare for the fire's arrival on Black Tuesday. This informal help suggests the role of civil society is critical to community safety but is often invisible in disaster management thinking and planning. A more diverse group at the emergency services table would be beneficial.

The expertise of volunteer brigade members is sometimes referred to in community education talks, when comments such as 'Why don't you go and have a talk to your local brigade?' (about local fuel issues, advice on purchasing pumps and hoses etc.) are made to the public.

- What are the pros and cons of this strategy from the public's point of view?

Community safety messages may also need to acknowledge the decision-making burden men (or is it 'society'?) place themselves under. South Australian research with women who wish to learn skills which have traditionally been considered male skills (such as firefighting) have expressed a preference to do this in a female-only learning environment (Long & Honner 2006).

- What principles would have to be implemented for this development to be carried out as a community development project? The writing of South Australian researcher Cheers (Cheers 2002) may assist with developing these ideas.

The relative absence of the 'home as refuge' idea

Despite the number of our households who were members of the CFS, only two interviewees spoke explicitly about the home as a refuge from the fire-front. One was a past member of the fire service (but not a farmer) and the other was a couple who are discussed in the next section. Some of those whom we have described as having 'bushfire experience' decided to shelter in their sheds rather than the house, and said they would do so again. In one case the shed was open on one side. Some people in this group were inclined to speak of cleared areas around a house or shed as areas in which to seek shelter. Sometimes the car was the 'shelter' associated with this idea of going to an open space. We did not push these discussions in our interviews, and there may be reasons for those decisions that were not apparent to us. Some aspects of the Australasian Fire Authorities Council policy of 'prepare, stay and defend or leave early' were not relevant to many farmers, for whom the presumption is already 'stay' (even if that presumption does not hold true for all members of their household). But there may be other beliefs which also mean that they do not differentiate between the 'home' and other farm dwellings in terms of taking shelter.

In two interviews (not involving individuals with fire experience) household members had drawn definite post-fire conclusions about not sheltering in the house. In one case, a husband was very regretful that he had asked his wife to leave the property as the fire approached. She drove into the smoke and crashed the car (but was unhurt), while her husband stayed at the property. The woman said that next time she wanted to stay inside the house, but her husband had argued that the best place for her would be in a car on the bare patch of ground outside the front fence.

Box 5.2: The home as refuge – implications

Some community members see farmers as having particular expertise in relation to response to fire. Yet the 'prepare, stay and defend or leave early' message may be targeted at a more urban-based population, with one main dwelling on their property. Facilitation of community groups may need to take into account the fact the behaviour of some community leaders, such as certain farmers, may be observed and emulated by others (see Rogers 1995 for these processes), yet the contexts in which decisions are taken may differ considerably. It may be necessary to be mindful of and tease out some of these differing contexts in group discussions in communities.

One of the dilemmas about the coronial process is that it ties up documents with scientific findings about many aspects of the fire, for far too long. Our experience showed that members of the public deliberate on their observations and what they hear in community discussion and draw conclusions from their local context of the fire event rather than from general findings from other events. The delay in releasing expert reports contributes to misinformation and the sometimes wrong conclusions drawn after these events.

Another household (father and adult daughter) came to the firm view that the house was not the place to shelter in. The daughter referred to 'those adverts on the TV that run all the time, "stay inside until the fire-front has passed"'. She added, 'But it didn't pan out like that at all'. We asked about their current understanding of that fire message. They both disagreed with it. The father added:

> In quite a few cases I could name ... people have in fact lost their houses either because they weren't there or they shielded inside the house and when they went out, the house was ... alight ... with burning in the ceiling and all this sort of thing which they weren't aware of.

We asked if they thought they saved the house because they were outside. One replied 'Definitely' and the other 'No question'.

The influence of children in thinking about preparedness and response

The role of women as carers of children appeared to be a factor in decision-making about who should stay with the property and who should leave. One young woman gave a clear example of this. She was pregnant at the time of the fire and was in town, while her husband was involved fighting the fire on private fire units. At the time of the interview, six months after the fire, her baby was four months old. When asked whether they would do things differently next time, the husband commented, 'I think my plan here would be the same, would be just stay in the house until the fire has passed and then go out and assess the damage because it offers some sort of protection'. This case is cited above as one of the exceptions where the household members spoke explicitly about the home as a refuge. We asked his wife if she felt the same. She answered that generally houses could offer shelter and a place from which to fight the fire after the front passed, and that:

> From the stories that you hear, the more people that left in cars, either were injured or died, than people that stayed in their houses. So, I think I would be more scared to be caught in a car than to be caught in a house.

We asked if that would be her thinking now that her circumstances had changed and she had a baby. She then took the conversation in a different direction and started to talk about the foreshore and the jetty, or 'calling Jack and jumping in his boat'. Then she brought the conversation full circle again and knowingly finished with 'I think I'd just stay with the house'. We think this brief conversational outline offers a microcosmic perspective of the paradox that houses are the safest place to shelter, yet when it comes to children there is an understandable tendency to think of removing them from the possibility of a home coming under ember or direct flame attack. In an area such as the Eyre Peninsula the neighbouring sea holds great appeal. This is both understandable and problematic, given the research and common experience of the danger of driving during a bushfire.

In three other interviews, it was assumed that if the children were at home at the time of a fire (which they were not in the households we interviewed) efforts would be made to 'get them to safety' and that if the woman was present that would be her role. In one household, the woman left the property and travelled through the fire in order to get to her children, who were in a nearby town. She had formed the view that her husband (who remained on the property) might not survive and that she too might not if she stayed, and that her children 'needed at least one parent'. A father tried to return from the fire-affected area to a nearby town where his family resided, when it seemed that the town might be hit by the fire.

While we can say that the presence of children is a key factor in household decision-making, our own data lacks input from parents who had babies and young children at home on the day of the Wangary fire. Proudley's (2007) research will explore this theme, in her analysis of interviews conducted in spring 2006 across 39 households on the Lower Eyre Peninsula, in greater depth.

> ### Box 5.3: Decision-making regarding children in fire response – implications
> Societal expectations about the role and responsibilities of the primary carer may overexpose women and children, in relation to fire, to dangerous outcomes. How might this issue be broached in discussion with community groups?

Conclusion

This chapter highlights the importance of experience with firefighting as a core aspect of community safety, and queries the relative absence of 'home as a refuge' idea expressed by those with such experience. It demonstrated the value of exploring the concepts of role and identity, gender and relationships, in the context of community safety in relation to the threat of bushfire. All these issues warrant more dialogue. This needs to occur at multiple levels and in 'safe' household, community and organisational environments. At the household level, members need to feel able to raise issues and expect their viewpoints will be heard respectfully. At the community level, members can explore with each other how they might take up new roles in working toward safer communities. At the intra and inter-organisational level, fire services and other relevant organisations and groups would benefit from meeting with households and communities to mutually reflect on what has happened, and how the roles and tasks of improving community safety can be shared and taken further.

Endnotes

1   We would like to acknowledge Alan Rhodes (CFA/RMIT/Bushfire CRC). He constructed and managed the telephone survey of fire-affected residents (Rhodes 2005), assisted our early learning in this complex field and conducted the majority of the interviews with us in July 2005.
2   The Australasian Fire Authorities Council provides advice to the public about what they should do during a bushfire. In this chapter we refer to the national policy position as the 'prepare, stay and defend or leave early' policy. The position is sometimes referred to as the 'stay and defend' or 'stay or go' position. The AFAC *Position paper on community safety and evacuation during bushfires* is available at: www.afac.com.au.
3   A more detailed analysis of aspects of this work is contained in Goodman, Healey and Boulet (2007).
4   The word 'dialogue' is used specifically as it involves frameworks for containing contentious and intrinsically emotionally charged issues (Issacs 1999).
5   Role as a concept arises largely from sociological literature (Turner 2002), where varying emphases mean that it can be prescribed and that it can be something people can step into and out of. It can also denote 'the meaning which the acts and symbols of actors in the process of interaction have for each other' (Conway 1978: 20). This idea overlaps with the idea of identity, which takes into account issues of social belonging and thus ideas of history and power relations, and issues of choices we make. Segerson (2002) draws on Sen's work and emphasises that choices can be understood as reasoned decisions reached from moral persuasions. Some of the data presented in this chapter shows that while safety messages might address individuals, 'individuals' can often be seen as operating from a more social, relational basis.
6   We acknowledge that a significant number of women have taken up operational rather than support roles in the fire services (Beatson & McLennan 2005).
7   What cultural values are attached to what roles? What changes in social and economic structures and values are we experiencing, and how do these changes affect our understanding of 'community' in 'community safety'? Who are the parties to the notion of 'shared responsibility' for community safety – shared responsibility between 'community' and 'government' – and what desirable practices might underpin such responsibility?
8   All interviewees have been assigned a pseudonym.
9   For general reading on the subject of gender and disaster see Enarson and Hearn Morrow (1998), Fothergill (1996) and Enarson and Meyreles (2004).

Assisting the Householder and Small Business Operator

Painting by Mark Schaller

Chapter 6

Prepare, stay and defend or leave early
Evidence for the Australian approach

Amalie Tibbits, John Handmer, Katharine Haynes, Tom Lowe and Joshua Whittaker

Chapter summary

This chapter provides an outline and overview of current research into the Australian 'prepare, stay and defend or leave early' policy. Although historical and structural evidence have been reported elsewhere, this chapter brings these issues together with additional verification from an analysis of a bushfire fatality database and findings from recent research into implementation.

The methodologies employed for this research include documentary analysis and extensive literature reviews, statistical analysis of bushfire fatalities, and numerous interviews, focus groups and quantitative surveys with members of the public and fire agencies before and after significant bushfire events.

This investigation demonstrates that the 'prepare, stay and defend or leave early' policy is very well-grounded in the available evidence. However, further research, outreach and community partnerships are needed to reinforce and clarify that leaving early or staying are not interchangeable options along a continuum which includes 'waiting to see' how the situation develops. In addition, clearly identifiable vulnerable groups need to be targeted with appropriate advice and, potentially, policy adaptations.

Introduction

Australian fire authorities believe that residents in bushfire-prone areas should be encouraged to decide, before the start of each fire season, whether they will prepare, stay and defend their property from bushfires or leave well before the fire arrives in their area. This advice is the basis of the 'Prepare, stay and defend or leave early' policy (referred to hereafter as 'the policy'), which is set out in the Australasian Fire Authorities Council (AFAC) *Position paper on bushfires and community safety* (2005) and endorsed by all Australian fire agencies.

The policy is underpinned by abundant evidence which shows that late evacuation is a dangerous response to bushfires that well-prepared houses can be successfully defended from bushfires and that they can provide safe refuge for people during the main passage of the fire front (Handmer & Tibbits 2005; Ch. 7 this volume). This evidence suggests that prepared but otherwise untrained people can protect their homes from bushfires by staying with and actively defending them. Since its inception, the policy has developed to become the centrepiece of community bushfire safety strategies in Australia (Gledhill 2003). Importantly, the Australian position is a significant move away from the evacuation doctrine that prevailed among emergency services in previous decades, towards greater community self-reliance.

This chapter has three main aims:
- provide an overview of the AFAC position and outline the main evidence supporting the policy from published material and a database of bushfire-related deaths
- provide evidence on the protection offered by structures
- describe impediments to the full implementation of the policy as identified by recent research.

The AFAC policy

Basics of prepare, stay and defend or leave early

The basic message of the AFAC position is that able-bodied people should be encouraged to stay with their homes. It states that:

> By extinguishing small initial ignitions, people of adequate mental, emotional and physical fitness, equipped with appropriate skills and basic resources, can save a building that would otherwise be lost in a fire (AFAC 2005: 6).

However, AFAC clarifies that this transfer of decision-making responsibility to the public should not be treated lightly and is subject to certain conditions. First, a key responsibility of fire agencies is to educate the public to develop the skills, knowledge and confidence they need to stay and defend their properties from bushfires. For example, people who decide that they will leave early are advised to decide where they will go, how they will get there and what their trigger for leaving will be. It is emphasised that leaving early does not necessarily mean going far, for example, a neighbour's property might provide safe refuge and enable quick return. People deciding to stay and defend should be physically and mentally prepared with firm plans to reinforce their commitment.

Second, although the AFAC position provides for mandatory evacuations in certain situations, there is an unequivocal statement that 'adequately prepared and resourced people should not be forcibly removed from adequately prepared properties' (2005, p. 8, emphasis added). In Victoria, legislation upholds residents' rights to decide whether they will stay and defend or leave early; that is, there is no legal provision for forced evacuation.[1] In other states, emergency services have the power to evacuate people who they believe to be in danger. Where legislation gives police the power to evacuate, AFAC recommends that a formal agreement be developed to ensure that such decisions are taken by the lead fire agency. The legal issues are explored in further detail by Loh (Ch. 8 this volume).

Finally, the AFAC position states that the most important aspect of preparation is the creation and maintenance of 'defendable space' – a space around the house in which fuels are reduced to protect against ember attack and radiant heat. Houses must also be well-prepared in order to provide safe refuge during the main passage of the fire front. Crucially, residents are advised to prepare their properties for bushfires regardless of whether they intend to stay and defend or leave early. This increases the safety of the property should a fire threaten without giving adequate time to leave, forcing the occupants to shelter in their home (CFA 2004). In any case, houses that are well-prepared but not actively defended – by residents or fire agencies – are more likely to survive a bushfire than those that are not (Wilson & Ferguson 1984; AFAC 2005). It is noted that there will be a small number of cases where the construction, materials and location of houses, including their proximity to unmanageable fuel loads, makes staying and defending impracticable. In such cases, residents are advised to leave early.

Historical evidence for the policy

Australia has had many bushfires resulting in large-scale destruction of homes and loss of life. In rural areas, 'staying' has always been a likely choice of survival strategy and in more remote locations it may be the only strategy available. However, prior to the 1983 Ash Wednesday bushfires in Victoria and South Australia, there were few formal investigations into the exact circumstances under which people died in bushfires, or the related issue of how houses are lost. An investigation after the 1967 Hobart fire, which resulted in 62 deaths and the loss of 1300 homes, showed that 'Most of the people who died in their homes ... were either very old and infirm or ... [had some] physical disability. In the case of about half of the people who died whilst escaping from their homes such homes did not catch fire' (McArthur & Cheney 1967). While the idea of staying with the house to survive a bushfire was clearly advocated in some circles prior to Ash Wednesday, the fires provided important evidence for more widespread acceptance.

The report of the post-Ash Wednesday Victorian Bushfire Review Committee identified a number of problems relating to evacuation and community safety. These included timely and accurate warnings, transport problems, evacuation routes and the need for clear definition of agency responsibilities (Miller et al. 1984: 27). The serious nature of the problems arising from late evacuation led the authors to recommend evacuation be considered a measure of last resort in bushfires, unless done early (which they state as the likely first preference). They also advised that attention should be focused on 'awareness and preparedness measures which assist persons to defend their own lives, homes and farms' (Miller et al. 1984: 137).

The Ash Wednesday Bushfire Review Committee also found that sheltering in a communal building or a house was a viable option that had saved lives in some cases. It qualified this by saying that people who survived in their homes generally knew what to do, had made preparations and, most importantly, had an adequate water supply. The committee acknowledged that during Ash Wednesday some people were apparently ordered to evacuate when they might 'reasonably' have been left to defend their homes. Concluding advice on this point is that 'In principle, people who choose to stay and defend their home or property should be allowed to do so' (Miller et al. 1984: 137).

In addition to the official inquiry of the Bushfire Review Committee, research by Wilson and Ferguson (1984) and Lazarus and Elley (1984), on communities just beyond the urban fringe, added to the argument against evacuating communities in the face of approaching bushfire – they also stressed the danger of last-minute evacuations.

The Dandenong Ranges fires of 1997 is apparently the only fire event to directly draw into question the 'safe to stay' doctrine. These fires killed three people, all of whom perished together while sheltering in a room in the house they believed to be fire-proof. The comments of family members and neighbours, documented in the coronial report (Vic. Coroner 1997), show a belief that the deaths could have been avoided if the three had chosen to evacuate when the fire approached their house. Some witness statements suggest that the CFA's 'safe to stay' message contributed to the deaths – an argument apparently strengthened by the fact that no neighbours perished in the fire, despite some late evacuations. This case stands out because, rather than contradicting the 'safe to stay' message, it highlights the importance of residents understanding what to do if they choose to stay. It also emphasises that staying is not simply passively 'sheltering in place' – it involves active defence. For a comprehensive review of the historical evidence for the AFAC position, refer to Handmer and Tibbits (2005).

Research into building ignition during bushfires supports the assertion that well-prepared houses can be successfully defended and can provide safe refuge during the main passage of the fire front (Leonard & McArthur 1999). It shows that wind-blown embers, rather than direct flame contact or radiant heat, are the most common source of house ignition before, during and after the main passage of the fire front (Leonard 2003; Ch. 7 this volume).

The earliest known published study concerning bushfires and houses was by Barrow in 1945. It demonstrated that houses predominantly ignited from the inside (e.g. under floorboards or in roof cavities) as a result of embers entering or becoming lodged around the house. Contrary to popular perception, houses did not simply 'explode' in the face of radiant heat from the fire. By showing that houses, including timber structures, were defendable if a few basic preparations were made, Barrow provided a basis from which others might argue that the home was potentially a safe place to shelter.

It became evident that residents could shelter inside well-prepared houses during the main fire front while extinguishing small ignitions in and around the home before and after the passage of the fire-front. For example, in their study of the 1983 Ash Wednesday bushfires in the Otway Ranges, Ramsay et al. (1987: 50) found that residents 'were able to save their houses by extinguishing small ignitions of the house itself before these fires became uncontrollable'. At Mount Macedon, also during the Ash Wednesday fires, a 90% survival rate was recorded for houses that were actively defended by able-bodied occupants, compared to 82% of attended (but not actively defended) and just 44% of unattended houses. The conclusion was that 'provided they are adequately informed of the danger and risks involved, mature, able-bodied residents can minimise loss of life, and probably save their houses, by staying within the safety of their homes' (Wilson & Ferguson 1984: 235). Accordingly, the AFAC position paper recommends that:

> With proper preparation, most buildings can be successfully defended from bushfire. People need to prepare their properties so that they can be defended when bushfire threatens. They need to plan to stay and defend them, or plan to leave early (2005: 5).

Bushfire fatality database

Haynes and Tibbits (forthcoming) verified and analysed a database of Australian bushfire fatalities. The database was originally compiled by Risk Frontiers[2] from information in the print media over the last 100 years and from various government reports. The database is a unique opportunity to assess the circumstances in which people perished. It provides details of each fatality, enabling a thorough analysis of people's actions as the fire front passed and the demographic relationships. In addition, unlike most other death datasets, the names and ages of individuals have been recorded, enabling verification with coronial reports. The value of coronial reports lies with the witness statements which allow an understanding of the stories behind decision-making. Most importantly, we can see that deaths cannot be simply attributed to age or physical ability. The database contains details of 689 fatalities, but as the investigation was concerned with the general public the deaths of the 123 firefighters were not studied, leaving a total of 566 fatalities. Figure 6.1 shows bushfire fatalities over the last 100 years.

Although the analysis is still at a preliminary stage, the data convincingly show the dangers of being caught outside during bushfire. The circumstances surrounding 327 of the deaths show 18% were inside a defendable property and 78% were outside or in an undefendable refuge (4% died in an unknown location). This is summarised in Figure 6.2. Of particular note is the significant proportion, 180 in total, of people who died while attempting to evacuate (this number includes those who chose to hide in an undefendable refuge such as a creek or culvert while evacuating).

Further investigation into the actions of the 18% (59) who died inside their property demonstrates that 8% (5) were actively defending their property, 58% (34) were not defending

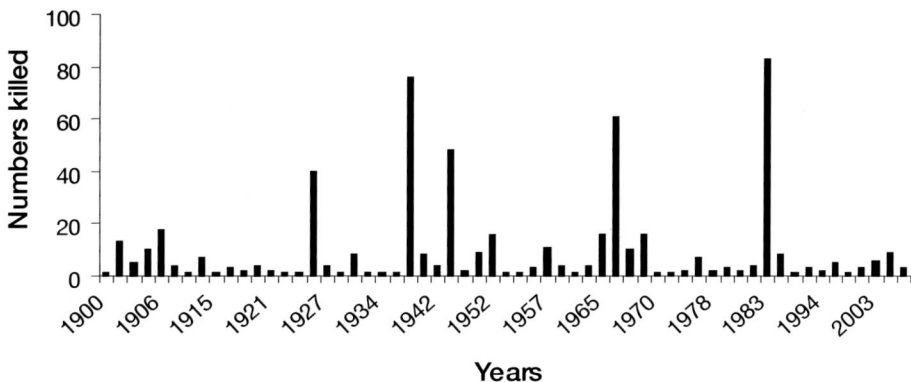

Figure 6.1   Australian bushfire fatalities 1901–2006. Peaks indicate mass casualties: 40 in Victoria (1926), 76 in Victoria (1939), 48 in Victoria (1944), 61 in Tasmania (1967) and 83 in Victoria/South Australia (1983).

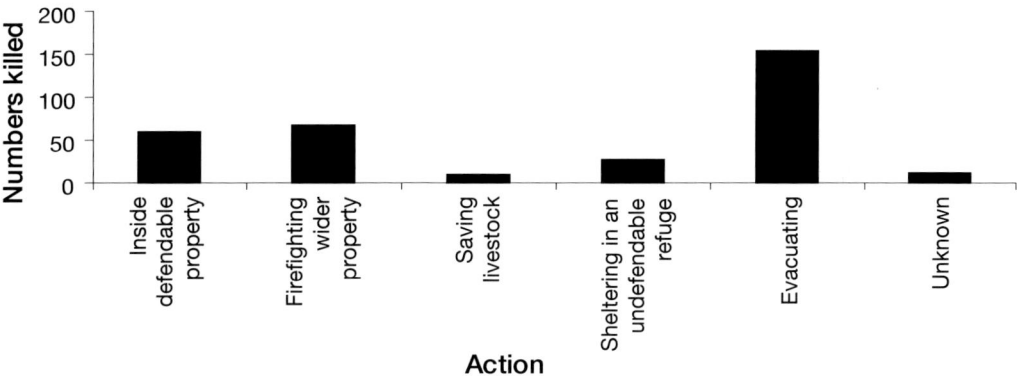

Figure 6.2   Actions undertaken at time of death.

and were sheltering and 34% (20) were in their home but their activities are unknown. This evidence confirms the importance of active defence rather than simply sheltering in place.

In addition to substantiating the evidence for the policy, the analysis found that most fire deaths are among men. However, the number of women dying in fires has increased over the past 30 years (Fig. 6.3). The relationship between gender and actions taken (Fig. 6.4) suggests that men are often caught outside, often defending their wider property (such as on farms – fences, sheds etc.), while most female fatalities occurred while sheltering in the house or attempting to flee.

This is also true for children and adolescents, with the majority of deaths occurring while attempting evacuation. The database lists 88 victims confirmed as under the age of 20, 56 of them under the age of 12. The actions at the time of death are known for 57 of these 88 victims (Fig. 6.5). Most notably, evacuation deaths in the past 40 years have shifted from predominately being on foot to late evacuation in motor vehicles (Fig. 6.6).

During the 2006 Eyre Peninsula fires, eight of the nine people killed, including four children and three adult females, died in or near their cars after attempting to flee (Deputy State Coroner 2005). Rather than being an anomaly, this event demonstrates the disturbing trend of the deadly actions carried out by a vulnerable group. For further detail of this fire, in particular the influence of gender on behaviour, see Chapter 5 this volume.

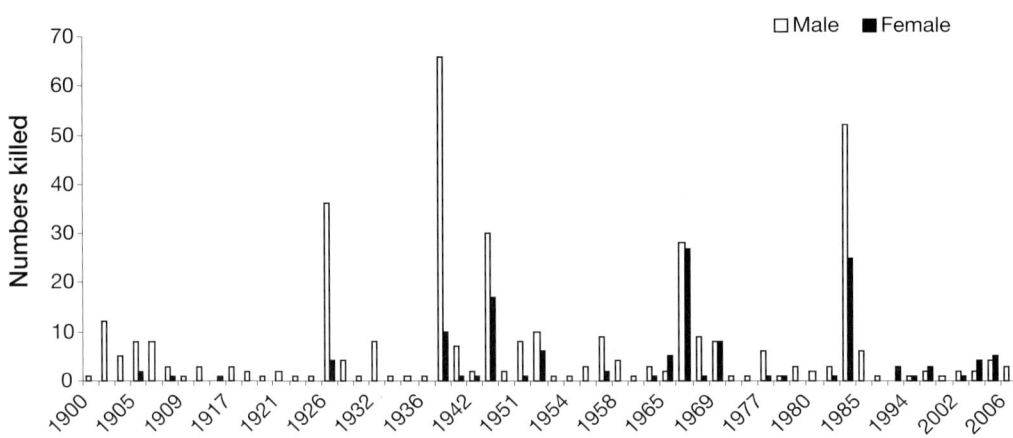

Figure 6.3 Gender distribution of deaths (388 males, 133 females, 45 unknown gender).

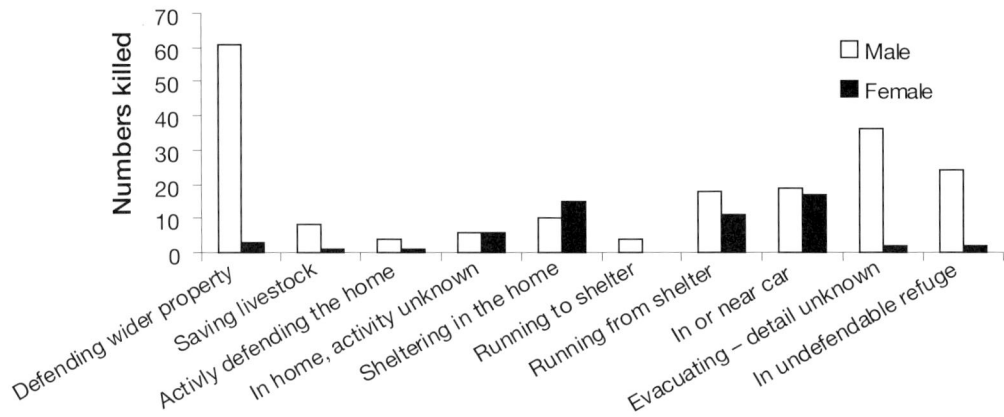

Figure 6.4 Relationship between gender and action taken at time of death.

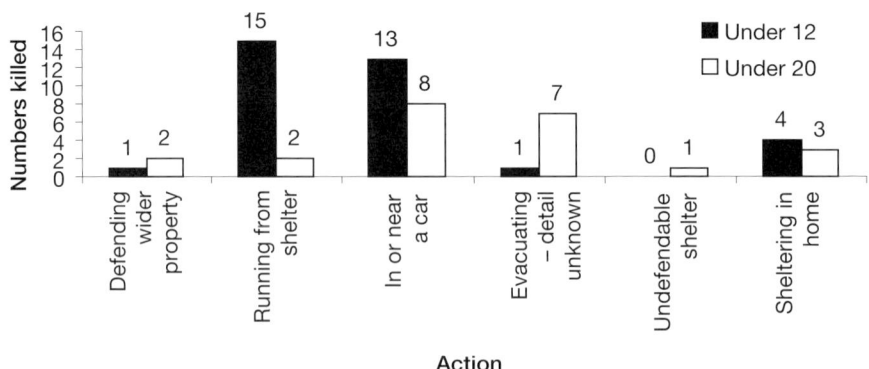

Figure 6.5 Child and adolescent actions at time of death.

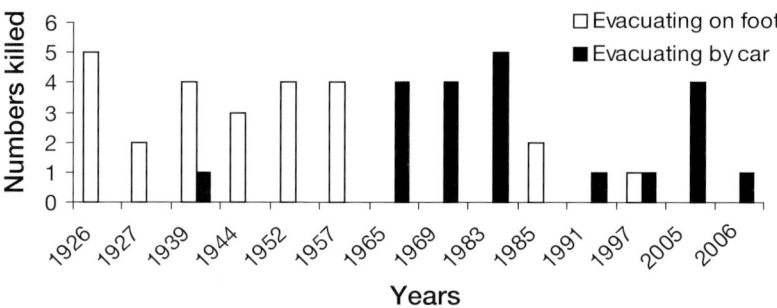

Figure 6.6   Trends in child fatalities during evacuation.

Analysis of this database is ongoing. With the addition of information from coronial records and collaborative work with CSIRO, subsequent analysis will allow researchers to further substantiate the circumstances in which people died and feed this information into improved policy implementation.

Case studies and implementation issues

The policy, while supported by a great deal of historical evidence, is difficult to translate into community action. The Centre for Risk and Community Safety, RMIT University, is researching several case studies to identify impediments to the full application of the policy in rural and urban interface communities. Case study sites are in Victoria, New South Wales, South Australia, the Australian Capital Territory and Tasmania.

This section will briefly outline the research conducted at those sites, then present the implementation issues emerging from the data.

North-east Victoria 2003

Methods and aims

Seventy-three participants attended five focus groups held in north-east Victoria and Gippsland after the 2003 bushfires (31 women and 42 men), in the townships of Beechworth, Bright, Mitta Mitta, Mount Beauty and Omeo. The focus group discussions revolved around six key questions designed to explore people's experiences of the 2003 bushfires and, more specifically, their experiences, understandings and opinions of the 'Stay and defend or leave early' policy.

Background and research highlights

From January to March 2003, bushfires swept across much of south-east Australia, burning 1.12 million hectares of public and private land in north-east Victoria and East Gippsland. Forty-one houses were destroyed and more than 11 000 head of stock were killed. Thousands of kilometres of fences were lost, as well as sheds, machinery and other agricultural equipment. The long duration of the 2003 fires – 59 days – and the fact that some communities were on alert for weeks before the fires hit them meant that residents had more time to prepare themselves and their properties than can normally be expected. Despite this extended preparation period and considerable access to information regarding 'prepare, stay and defend or leave early', some participants revealed serious misinterpretations of the message. In particular, some people had decided to leave early but were relying on a trigger such as advice from authorities or the appearance of smoke and flames. Of those planning to stay and defend, many consciously or subconsciously retained contingency plans in case they felt unable to defend their property from fire. Most commonly, and despite widespread recognition that late evacuation is extremely dangerous, some of those who stayed and defended had their cars packed and were prepared to leave at the last minute if they felt it was necessary. Others had identified a nearby place to which they could retreat if they felt threatened, such as an open area, cellar or mineshaft.

The discussions revealed a high level of awareness of and support for the policy. Participants firmly upheld their right to stay and defend their properties and the vast majority were vigorously opposed to ordered (or forced) evacuation (see Tibbits & Whittaker (2007) for more detail).

Hobart 2006

Methods and aims

A month after fires bordered suburbs of Hobart's eastern shore (October 2006), 49 face-to-face surveys were administered with residents in the affected areas. Questions centred on preparedness, decision-making, information sources and expectations of fire services during the incident.

Background and research highlights

11 October 2006 was a day of extreme fire danger, worse than the day of the devastating 1967 fires. Tasmania Fire Service (TFS) resources were already stretched across several blazes when a fire started in bush adjacent to the eastern shore suburbs of Hobart. Despite extreme fire weather and the fire's proximity to a residential area, no houses were lost and damage was minimal.

All respondents surveyed had stayed to defend their property, many doing so on the advice of the TFS, police or local ABC radio. The TFS utilised its resources effectively, moving units ahead of the fire front so appliances were always at residential areas when the fire front hit, giving residents the impression that they were not alone. The presence of the TFS and having a fire appliance in the neighbourhood increased the confidence of residents to stay and defend. Some people were given advice directly by firefighters, such as the most effective use of water or to put on protective clothing. Some also stated that firefighters had been very encouraging of their efforts to stay and defend, yelling things such as 'you're doing a great job, keep it up' from the truck as they drove by.

Despite a very high level of confidence about staying and defending their property, the study highlighted residents' reliance on fire services for advice and psychological support. Residents did not seem to have the complete self-sufficiency assumed to be required for successful application of the 'stay and defend or leave early' policy. Many interviewees stated they did not feel threatened because there was a fire truck near their property, with several noting they would have left if they had felt threatened. Residents expressed very high expectations that the TFS would protect property and provide warnings. See Lowe et al. (in press) for more detail.

Canberra 2003

Methods and aims

Nineteen in-depth interviews were conducted jointly with CSIRO in Chapman, a suburb heavily hit by the 2003 Canberra fires. The interviews provided a narrative of people's experience on the day and the factors which influenced their decision to prepare, stay and defend or evacuate.

Background and research highlights

In January 2003, bushfires in the western suburbs of Canberra resulted in four deaths, the loss of over 500 houses, extensive infrastructure damage and the loss of much commercial forest and the southern hemisphere's leading astronomical observatory. The day the fire hit the city was one of extreme fire weather.

Despite little warning, a low perceived threat of bushfire and a lack of pre-fire preparedness, many people did successfully stay and defend their property. Residents in Chapman revealed they had little more than 30 minutes to prepare before their properties were hit by a severe ember attack and fire front. In that time they were able to make basic preparations, such as filling containers with water, blocking and filling gutters with water, clearing flammable fuel from around the house and sealing the house against embers. Many cited local ABC radio and neighbours as key information sources for necessary preparations. This last-minute preparedness, followed by vigilance in putting out small fires around their properties, was instrumental in saving many properties that may otherwise have been lost.

People's preparedness and active defence did not always ensure house survival, with several of the interviewees losing their property. The house did, however, provide a safe haven while the fire front passed, and staying and defending prevented some people from evacuating before or during the fire front. By the time houses were well alight beyond recovery, the fire front had passed and it was safer for people to leave the structure.

The New South Wales and South Australian studies are described in detail in Chapter 3 and Chapter 5 this volume, respectively.

Discussion of the results

Knowledge of the policy

On the whole, our research has found that residents in fire-affected areas have a high level of awareness of the policy; this is to be expected due to their frequent and recent bushfire experience. Despite this awareness, the comprehension and translation of the policy into an appropriate bushfire response has been varied. Of particular note are the following issues:

- a lack of pre-season household planning and preparedness
- 'prepare, stay and defend' is often interpreted as 'stay and defend until I feel threatened' or 'wait and see'
- a lack of clarity about what 'leave early' actually means, including when to leave and where to go
- expectations that fire services will be available to assist in property protection, give property-specific information, provide individual warnings and provide evacuation orders.

Issues surrounding education and outreach

Little is known about the effectiveness of bushfire education and outreach activities (Ch. 13 this volume). However, a growing body of bushfire research and more general hazards research has shown that educational warnings frequently fail due to fundamental differences in the ways expert practitioners and members of the public view and make decisions about risks (Ch. 2 this volume). The judgments of emergency managers are often based on formal assessments, for example, estimates of the probability of a hazardous event and the likely number of fatalities. In comparison, judgments made by the public are based upon more subjective ideas which can be heavily influenced by people's beliefs, attitudes and feelings as well as their wider social and cultural values (Morgan et al. 2002; Ch. 11 this volume).

While the differences in judgment are not vastly dissimilar, it is clear that giving people the knowledge that experts think they need is no guarantee that people will act in the ways that emergency mangers want them to (Handmer 2000). The idea of 'resilience' therefore extends far beyond people simply knowing what to expect and how to react when threatened. Instead, our research suggests that partnerships should be formed with communities in an effort to accommodate all levels of risk perception and understanding (see Ch. 4 and Ch. 3 this volume). However, this is a problematic task that requires trust, shared responsibility, co-operation and communication across numerous organisations, professionals and community groups. A major challenge in fire-prone areas is tapping into existing sources of social capital or creating new bonds within high-risk communities. This is of particular importance when dealing with issues surrounding the policy and has significant implications for rethinking how we communicate and control risks.

Deciding to stay and defend or leave early

Decisions to stay and defend or leave early are often complicated by a range of factors beyond concern for personal safety or property protection. People's confidence in the survivability of their house and their own physical and mental ability to stay to defend it are fundamental. Other considerations, however, such as responsibilities for vulnerable household members, level of commitment to property, work and emergency-related responsibilities are also factored into decision-making and planning. It is important to acknowledge how individual circum-stances may conflict with or impede a person's or household's ability to make and act on the policy as recommended by fire authorities.

Commitment to stay/contingency planning

A key element to successful defence is having the commitment to stay throughout the fire and not adopt an approach of waiting or staying until there is a feeling that the bushfire threatens personal safety (as is most common). Brennan (1998: 11), in her study following a fire in Macedon, Victoria, described full commitment as 'a decision made prior to the incident, where no actions of any household member contradicted the plan and where there was no wavering when they were asked about their response to a scenario where firefighters were absent and fire was approaching'.

Brennan described several features that distinguish households committed to staying from those that are not. These include having a plan developed and understood by all household members, not leaving the premises to seek advice or information or to watch for the fire, and identifying defence as a system, not just the application of a few suggestions.

Our case studies found that many of those who plan to stay and defend their properties are not fully committed to doing so. Many who decide to stay and defend are consciously or unconsciously retaining late evacuation at the last minute as an option, despite widespread recognition of its dangers (Tibbits & Whittaker 2007). Some people consider that having a contingency plan, such as ensuring the keys are in the car and that the car is facing toward an escape route, is rational planning in case things go wrong or they feel too scared. Evidence that people plan to stay and defend but are prepared to leave their property if they feel threatened suggests a misinterpretation of the policy, which fire agencies must work hard to rectify.

Preparedness

Most of the problems people faced when staying to defend their properties result from poor planning. For example, many people do not take into account the likely interruptions to mains electricity and water supply. The cost of a water tank or a petrol-powered water pump may be prohibitive, but there are lower or no-cost alternatives, such as 44 gallon drums and rubbish bins filled with water (CFA 2006). In any case, people who are unable to actively defend their property from fire, for whatever reason, should plan to leave early.

Clarification of 'leave early'

There is significant evidence (Rhodes 2005a; Tibbits & Whittaker 2007) that the 'leave early' message is not well understood and requires clarification. Our research has revealed the decision about leaving early is often not made prior to the beginning of the fire season, as is advised by AFAC. Furthermore, the trigger to leave is often advice from authorities (personally or via the radio) or, in the worst cases, the presence of heavy smoke or flames in the immediate area. Although contingencies and individual circumstances preclude fire authorities from offering uniform advice on when to leave, it may be helpful to suggest triggers for leaving early that people can plan for in advance. It is critical that residents are able to identify appropriate times to leave, which includes knowing when it is too late and therefore dangerous to leave their property. In these circumstances, residents need to be aware that staying in a house, whether their own or a neighbour's, is much safer than attempting to flee in a car or by foot. This option is especially appropriate for people who are not confident to stay in their house alone or those for whom leaving early is impracticable (because of their remote location or the long duration of the fire).

Reliance on emergency services

There is still a widespread belief in the Australian community that emergency services will be available to assist residents if their property is threatened by bushfire and advise them when and where to go if they need to evacuate. Although this belief is contrary to the self-reliance requirements of the policy, it is consistent with the traditional roles of emergency services, such as co-ordinating evacuations. There is evidence to suggest that fire services and the media may be reinforcing notions of dependence through avenues such as: the Standard Emergency Warning System (SEWS), individuals (often volunteers) on the fire ground assuring residents they will receive a warning to evacuate, police ordering residents to evacuate, personal assistance from a fire unit, and stories in newspapers, radio and other media.

Trauma

Successfully staying and defending a property can be a positive experience. However, Odgers and Rhodes (2003) and Rhodes (2005b) noted that trauma and depression are often experienced by people affected by a fire event, particularly those who had to actively defend their property. The emotional cost of staying and defending needs to be weighed against the increased chance of house loss if residents choose to evacuate. Those who flee and survive despite driving through flames are also likely to experience significant trauma.

The psychosocial effects of bushfire trauma can be far-reaching and long-lasting, negatively affecting livelihoods, family relationships and general health (Mental Health Research and Evaluation Centre 1985). There is no doubt that experiencing a disaster is traumatic, but changing the emphasis of bushfires from 'terrifying and life-threatening' to 'dangerous but survivable' through residents' active participation may lessen the trauma. Many research participants spoke of feeling empowered by successfully defending their property and there is evidence that the experience can draw communities together. However, this is not always the case, particularly where the level of loss in surrounding areas and property has been significant. This is especially true in rural areas where homes may be less significant than other assets such as fencing, sheds, farm machinery, stock and feed.

A key recommendation of the COAG report following the 2003 bushfires was that the Australian attitude to bushfire should be changed. Shifting our notion of bushfire is particularly relevant when considering staying and defending, where last-minute evacuation continues to be considered as an option if people feel threatened. The trigger to leave is likely to be the sight or sound of the fire. Increasing the understanding of what to expect in a bushfire may decrease the likelihood of fear causing late evacuation. Furthermore, making 'staying and defending' the normal response to bushfire, rather than heroic and dangerous, is likely to reduce the associated trauma.

Conclusions

The 'Prepare, stay and defend, or leave early' policy is well supported by published evidence, our case studies, and an examination of bushfire-related deaths. It is also supported by other chapters in this volume that examine the legal issues, building safety and some of the case studies in detail. Empowering and encouraging people to decide whether they will prepare, stay and defend their properties from bushfires or leave early – provided they act on their decision – is probably the single most important strategy for protecting people and property from bushfires. However implementation of the policy could be improved. We set out issues for practice and for research with a view to improved implementation below.

Issues for practice

These are essentially about improved implementation.
- Clarification of 'Stay and defend' for property occupiers. Residents must understand that staying to defend their property requires a firm decision and a commitment to stick to their plans. Late evacuation is not an option.
- Clarification of 'Leave early'. Residents must understand what 'leave early' means – when and where to go – and must be able to recognise the point at which it has become too late to evacuate. Late evacuation is to be avoided.
- Greater emphasis on the need for residents who intend to leave early to prepare their properties for bushfires, both to increase its chance of survival and so that the property can provide safe refuge if residents miss the opportunity to leave early.
- What are the physical prerequisites for staying and defending? Mental preparedness is unknown at present: how can people prepare themselves psychologically?
- There is a need to work more closely with firefighting volunteers to ensure that residents have realistic expectations of support.
- Help for those who cannot or do not prepare.

Future research

- Research to support the implementation issues identified above under 'Issues for practice' as well as other broader institutional issues, for example police/firefighter relationship issues.
- Given the trends in climate and demographics, what gaps and issues are likely to emerge with the application and implementation of the 'Stay and defend or leave early' policy in rural and urban interface areas in Australia?
- Research could usefully examine the applicability of the policy to other countries, both to assist them with bushfire safety policy and to help illustrate other issues with the policy in Australia.

Endnotes

1   Please see s30(1)(g) and s31(3)(b) of the *Country Fire Authority Act 1958* (Vic.), s58 of the *Metropolitan Fire Brigade Act 1958* (Vic.) and s 24 and s36B of the *Emergency Management Act 1986* (Vic.). These provisions state that people may be removed only if they do not have a pecuniary interest in the land, building or goods.

2   Risk Frontiers is a not-for-profit research organisation based at Macquarie University in Sydney, funded largely by the reinsurance industry to help better estimate the cost of damage from natural hazards.

Figure 3.4 Bushfire CRC researcher Tom Lowe speaks to a CFU member at a 'fixed hose post' CFU in South Turramurra, Sydney, in October 2006.
Photograph: Katharine Haynes

Chapter 6 Fire-damaged car, burnt for experimental research conducted by the Bushfire CRC at the RFS Hot Fire Facility in Mogo, NSW.
Photo © Bushfire CRC.

Chapter 14 Erikson Aircrane at work in Gippsland, Victoria, January 2007.
Photo © Greg McCarthy, Bushfire CRC.

Figure 7.2 Examples of ember attack on decking (extinguished by resident).
Source: Blanchi et al. 2006a.

Figure 7.3 Examples of ember attack on eaves (left) and fascia (right).
Source: Blanchi et al. 2006a.

Figure 7.4 Window cracking caused by radiant heat from adjacent burning car (left) and window breakage as a result of frame ignition (right).
Source: Blanchi et al. 2006a.

Figure 7.5 Impact of surrounding burning vegetation on window (replaced with wood panel) (left) and the hazardous effect of pencil pines (right; arrow indicates row of pencil pines that ignited, causing fire spread to adjacent houses).
Source: Blanchi et al. 2006a.

Figure 7.6 Behaviour of fencing in bushfire.
Source: Blanchi et al. 2006a.

Chapter 7

Property safety
Judging structural safety

Raphaele Blanchi and Justin Leonard

Chapter summary

Bushfire is an inevitable part of living at the rural–urban interface, but a well-prepared house can offer a safe refuge, providing a better chance of survival than does a late evacuation. Also, able residents can extinguish ignitions before and after a fire front passes, greatly enhancing the chance of house survival. This chapter presents recent research into the area of building survivability, to support the development of regulations, planning policies and community education initiatives. Methods included surveys after major bushfire events and experimental procedures under simulated fire conditions.

In Australia, 'house loss' in bushfires encompasses houses destroyed without being directly affected by the flames and/or radiant heat from the main fire front. This presents many opportunities for effective risk mitigation, through an understanding of the specific aspects of urban design and the human behaviours that influence the potential for house survival. House loss follows a weakest-link principle, and effective and efficient approaches to bushfire risk management involve systematically identifying and eliminating the weakest points, until a suitable level of residual risk is achieved.

Virtually all house losses occur during extreme fire weather conditions. Therefore, these extreme conditions and their potential return period for a given region should be used as the basis for preparing for future events.

The key to effective bushfire risk management is the integration of various strategies to ensure an ongoing understanding of the risks, and to maintain the intended urban design outcomes. Such an approach would involve integrating the understanding of:

- hazard conditions (characteristics of bushfire events)
- house design and materials (from voluntary or prescriptive specification)
- the interactions between objects in the landscape
- the influence of brigade and occupant behaviour before, during and after a bushfire event.

If residents clearly understand the intent of these features and have general bushfire knowledge it will greatly improve their chances of survival.

Introduction

The desire to live at the rural–urban interface carries the responsibility to manage the risks associated with the arrival of an unplanned bushfire. It is clear that understanding the risks is essential in reducing life and property loss, however, while many people accept these risks and undertake measures to achieve an acceptable level for their own circumstances, others are ignorant of or ignore such risks. As a result, a minimum set of requirements is often prescribed in an attempt to reduce the risks in this demographic. Fire services and governments have issued several recommendations related to design and construction of buildings in bushfire-prone areas (e.g. Standards Australia 1999), as well as planning guidance (e.g. Fire and Emergency Services Authority of Western Australia (FESA) 2001; NSW Rural Fire Service 2006). The effectiveness of these prescriptions is greatly reduced if residents are unaware of their value, intent and required upkeep.

Communities act to reduce the consequences of bushfires when they occur. A well-prepared house is considered a safe shelter during the passage of the main bushfire front. The house may experience radiant heat and flame attack for a period of 10–20 minutes (the amount of heavy fuel load around a house also influences the time taken for that area to become tenable after the main fire front has passed), during which human activity outside the house would not be safe. Different studies have suggested that people sheltering in their house and implementing different protection strategies before, during and after a fire front have better chance of survival than people evacuating late (Wilson & Ferguson 1984; Krusel & Petris 1992). In their *Position paper on bushfires and community safety*, the Australasian Fire Authorities Council (AFAC) (2005) has endorsed the fact that a well-prepared home or evacuating/relocating well in advance of a fire threat are the best survival options during a bushfire (see Ch. 6 this volume).

Staying with a house also increases its chance of survival if the resident remains active in and around the house when it is safe to do so. After the passage of the fire front, able residents can monitor and suppress small ignitions in or around the house before they become uncontrollable. Previous research has shown that active defence of houses by residents or brigade members significantly increases the chances of house survival (Wilson & Ferguson 1986; Ramsay et al. 1986; Leonard & Bowditch 2003; Blanchi et al. 2006a).

This chapter presents recent research into the area of building survivability, to support the development of regulations, planning policies and community education initiatives. The first part of the chapter describes the methodology used to collect information on house vulnerability. The second part discusses the mechanisms of bushfire attack at the urban interface. Later sections focus on the vulnerability of houses (house design and materials), the interactions between different combustible and non-combustible objects (including vegetation) in a landscape and the building envelope that could contribute to or mitigate eventual structural loss, and the influence of human behaviour before, during and after a bushfire event.

Methodology

To gain a better understanding of the issues involved in loss as a result of bushfires, two approaches have been taken: collecting information via surveys after major bushfire events, and experimenting with the performance of various structural elements under simulated bushfire conditions.

Post-bushfire surveys

To increase our knowledge of the mechanisms involved in house destruction, several in-depth surveys involving over 2000 houses were conducted following large bushfire events that resulted in significant house loss. The first survey was conducted after the 1983 Ash Wednesday fires in Victoria and South Australia, and the latest was in 2005 after the fires in South Australia.

Each survey used a common approach, which has undergone a process of continual improvement since the initial 1983 survey, but with aspects of some standard questions maintained to provide statistical continuity. The survey form is predominantly focused on assessing building design attributes with a view to correlating these with the probability of house survival (McArthur 1997).

Data collected from on-site inspections, owner/occupant interviews and the records of shire councils and various organisations includes:

- details of the extent of damage (destroyed, damaged or untouched)
- extensive information on the structure of houses
- site details (e.g. land slope)
- description of surroundings
- details of the actions of residents and firefighters during and after the event.

Each of these elements are considered, specifically how they interact and contribute to a community's bushfire risk.

The information is stored in spatial navigable databases constructed in a standardised format to allow future cross-organisation data-sharing.

Laboratory and full-scale simulations

In addition to the information provided by the surveys, the results of experimental work have been used to better understand the performance of various building components subjected to bushfire attack and to develop methods for the design of these elements.

In the laboratory experiments, the performance of different material was tested using a cone calorimeter, or by adapting or designing specific experimental apparatus for a specific exposure condition.

For the full-scale simulations, a bushfire front simulator was used to recreate 'real' bushfire flame characteristics (flame temperature and radiant heat flux). Four types of bushfire exposure were performed:

- leaf litter exposure – simulated ember attack by the distribution of ground fuel litter
- bushfire pre-radiation exposure – ember attack and radiation simulating an advancing fire front that will not reach the element tested
- bushfire flame immersion exposure – ember, radiation and flame attack simulating an advancing fire front that burns through continuous forest fuels up to the element tested
- structure fire exposure – this level of exposure was designed to simulate a worst-case structural fire exposure for a period of 30 minutes.

Understanding and assessment of bushfire threat

Bushfire attack mechanisms are measured in terms of the heat and intensity of both the flame front and the flux of embers (burning debris and windborne debris) before, during and after the fire front. An assessment of bushfire threat is based on the amount and characteristic of bushfire fuels, weather conditions and land slopes. Also, extreme bushfire weather conditions can be considered as influencing the magnitude of a bushfire impact and the vulnerability of a structure and surrounding elements (materials becoming dryer and more flammable).

The relationship between house loss rate per single fire event, and the localised weather conditions under which this house loss occurred, has been analysed by Blanchi et al. (2006b). The results show that house loss rates greater than 50 occur at forest fire danger index levels of 40–60 (extreme), and that major house loss events (more than 100 houses) occur at index levels of 60–100. House losses in the thousands are possible when the forest fire index exceeds 100.

From the human perspective, the main threat to personal safety is radiant heat or direct flame contact, and it is essential that people take refuge within structures well before any significant flame front arrives. The radiant heat from an approaching flame front may be sufficient to cause serious injury or death and it is highly dependent on the proximity and intensity of the front. The first sign of untenable conditions is the onset of pain on bare skin. This occurs at approximately 2 kW/m^2, which can be exceeded when a flame front is more than 100 m away.

Once the fire front has passed, surrounding burning elements (sheds, stored materials or adjacent homes) may be significant sources of radiant heat. Therefore, building design should include sufficient clearance between a house and surrounding elements that it is possible to leave and/or defend the house after the fire front.

From the building perspective, ember attack has been identified as the main cause of house damage. CSIRO research showed that the majority of houses lost in bushfires survived the passage of the fire front, and burnt down during the following few hours due to lower-intensity fires spreading from ignition by burning debris (Leonard 2003).

The survey conducted after the 2003 Canberra fire identified a high percentage (>90%) of houses as damaged or destroyed in the absence of direct radiant heat and flames from the fire front itself (see Fig. 7.1). The predominant causes were ember attack, or radiant heat or flames from surrounding burning objects, leading to house ignition (Leonard & Blanchi 2005).

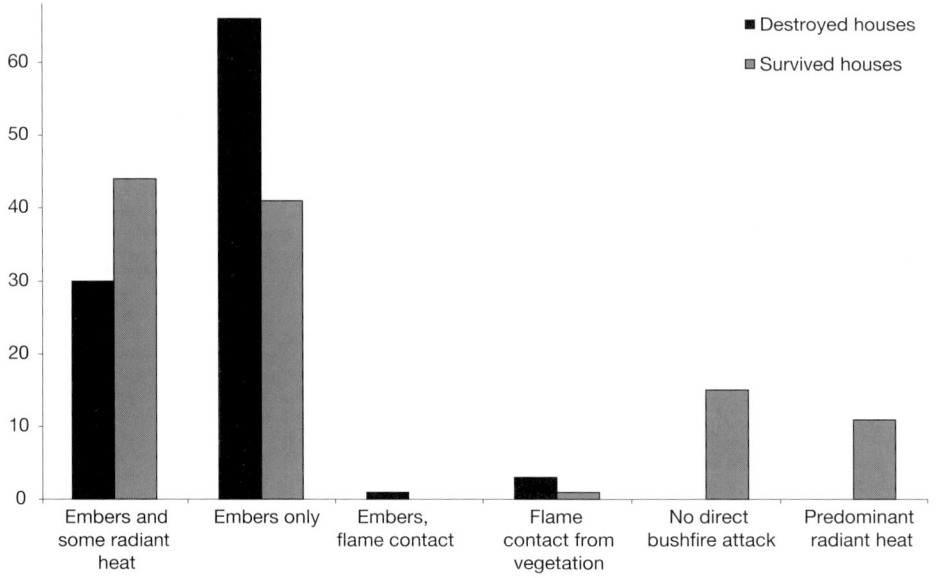

Figure 7.1 Mechanism of bushfire attack for surveyed area of Duffy (Canberra fire, 2003), expressed as a number of houses destroyed and survived ('survived' includes houses damaged and untouched – main building is untouched but surroundings are affected).
Source: Blanchi et al. (2006a).

Embers have proved to be a major cause of ignition, as they can attack a building for some time before a fire front arrives, during the passage of the fire front and for several hours after the fire has passed. Two types of windborne debris need to be considered: burning debris that can enter into or directly ignite parts of a house or its surroundings; and unburnt or partly-burnt debris that could facilitate ignition when it accumulates in or on specific parts of a house or surrounding elements (Leonard & Bowditch 2003).

Once an initial ember attack on a house has developed into a structural fire, an urban spread mechanism can develop in which burning houses facilitate direct flame, radiation and ember attack on surrounding structures. For example, during the 2003 Canberra fire, approximately 35 houses in the suburb of Duffy were damaged or destroyed by flame and radiation from adjacent houses (16% of the whole population) (Leonard et al. 2004), and many more were lost through indirect ignition or embers from other structures or urban vegetation.

The combination of embers and radiation caused by urban fuel combustion (adjacent structures, fences, stored material etc.) has been identified as the predominant ignition mechanism. Direct flame and radiation attack from the flame front itself plays a minor role in all bushfires, either because the flame front is not of sufficient intensity when it reaches a structure or because it cannot get close enough to a structure, even in areas without deliberate planning provisions. Surveys showed that for 'first-row' houses in the Otway, Sydney and Canberra bushfires, a combination of adequate setback distance and deliberate or accidental urban edge treatment was sufficient to decrease flame and radiation impact from the fire fronts. This observation suggests that urban edge treatment such as fuel reduction (e.g. firewood collection) or ground fuel management for increased access can play a significant role in reducing fire intensity at the interface.

Understanding house vulnerability

A house can be considered as an envelope, any breach of which could lead to house destruction. The major bushfire attack mechanisms are ember entry, ember accumulation, radiant heat and flames. The structural performance of a house under each individual mechanism has to be studied to understand the overall vulnerability of the house. Some aspects can be considered individually and others synergistically.

Ember entry through common gaps and entry points

In this case, vulnerability is linked to the possibility of ember entry through gaps in the structure. Gaps of more than 2 mm in the building envelope are considered potential points of ember entry (Standards Australia 1999). Different parts of the structure can be defined according to their likelihood to ignite due to the amount of combustible materials present, occupiable space, roof, subfloor and wall cavities (from most likely to ignite to least likely to ignite, respectively) (Blanchi et al. 2006a). Metal flywire screening is an effective method of protecting these gaps.

Ember accumulation

Vulnerability to ember accumulation against the building envelope depends on structural configuration and design. For example, re-entrant corners and crevices enable embers to accumulate, generating sufficient localised flame for adjacent combustible materials to ignite. The phenomenon of re-radiation between surfaces increases the opportunity for a small flame to develop into a significant flaming attack on a structure. Another parameter is related to the combustibility of construction materials.

Radiant heat from surroundings

Radiant heat and flames present a risk based on the level of radiant heat exposure, as well as the time over which the exposure occurs (Leonard et al. 2004). The vulnerability of a material also depends on its propensity to support localised flame development. Any combustible materials stored nearby also have the potential to ignite and threaten the house envelope.

Examples of vulnerable building parts

The results from post-bushfire surveys have shown that timber decking is the feature most likely to be directly ignited by embers, followed by eaves and gutters, and timber window frames. Ember entry into roof cavities and subfloor spaces is less prevalent (for the damaged house category), but residents are less likely to identify this attack mode until it has developed to a point where it is difficult to extinguish. Hence, this attack mode should also be considered significant even though there are few recorded examples of subfloor or roof cavity ignitions leading to total house loss (Leonard & Blanchi 2005).

Timber decking

If left unattended, small ignitions on timber decking or verandas can grow to a point where they threaten building elements such as facades, windows and doors, thus threatening the building envelope (see Fig. 7.2, page 74). Research into timber decking has shown that replacing the decking boards closest to the structure with non-combustible alternatives greatly reduces the potential for facade ignition from a decking fire. Eliminating fuel build-up or materials stored under a decking greatly reduces the intensity with which the deck may burn. Using non-combustible supporting members also significantly reduces the potential of ignition of decking. During windy conditions, flames from burning decking can be driven through very small gaps in an adjacent building façade, causing structural ignition. Attention to facade design adjacent to timber decking is essential.

Roofs, eaves and fascias

Roof valleys are key areas where embers and windborne debris can accumulate. The surveys showed that roofs with multiple ridges and gullies fare worse than flat roofs. Eaves and fascias have also been identified as key points of ignition and entry (see Fig. 7.3, page 75).

Glazing systems

Although windows are not directly affected by ember attack unless they are unscreened, open or poorly fitted, the glass can be easily damaged by radiant heat or flame contact from the fire front itself or from the combustion of building elements (e.g. combustible window frames) or nearby fuel sources (e.g. vegetation, decking, garage, shed, car, neighbouring houses) (see Fig. 7.4, page 75). If glazing cracks and falls away, windborne embers/debris are highly likely to enter and ignite internal furnishings.

Experiments (Bowditch et al. 2005) have shown that plain glass windows in timber frames can withstand radiant heat loads up to 12 kW/m^2, while common aluminium framed windows can withstand loads of around 14 kW/m^2. If metal flyscreens are used over the glass and framed area, these heat load tolerances increase by around a third. For toughened glass, the threshold for window failure is $30-40 \text{ kW/m}^2$ depending on the framing material (timber, common aluminium or aluminium with high-temperature seals).

External doors

External doors can present a particular hazard for ember entry if accidentally left open, although doors are less likely to be left open or ajar than windows are. External doorways can be a point of ember accumulation or provide a re-entrant corner with re-radiation effects from localised combustion of garden materials. The use of a non-combustible screen door or the use of low- or non-combustible materials in the door assembly is highly recommended.

Effects of surroundings on house loss

The surrounding environment (type of vegetation, fences etc.) and any outbuildings (type, materials of construction, proximity to houses) can increase or reduce the risk of house loss.

Surrounding vegetation

In the Otways (1983 survey) and Sydney (1994 survey), the amount of vegetation around a house was found to influence house survival, with thicker surrounding vegetation and a higher proportion of trees than shrubs more likely to result in house destruction (Ramsay et al. 1994). Overhanging trees appear to increase the risk, mainly because they deposit material on and immediately around the structure (Leonard & Blanchi 2005).

During extreme fire weather conditions, the vegetation immediately around a structure is likely to be very dry and vulnerable. These conditions and the type, amount and distribution of vegetation can support the progress of ground-based fire spread deep within an urban area, increasing the number of structures exposed to these fire effects (see Fig. 7.5, page 76).

Managing vegetation involves selective fuel reduction (removal, thinning, pruning) and the retention of beneficial vegetation that may act as windbreaks and radiant heat barriers (Leonard & Blanchi 2005). It is desirable to have a fuel-reduced area around a building to reduce the level of attack by flame contact and radiant heat. The extent of the fuel-reduced area depends on the type of vegetation, the slope of the land and its aspect. Fire authorities can give advice about the size of the area. Isolated trees that are not identified as highly flammable are acceptable within the region, and in many cases can offer a welcome windbreak and radiant heat barrier.

Outbuildings

The risk of outbuildings depends on their proximity to the home, their design and size and material within the outbuilding. Post-bushfire surveys show that outbuildings are more readily lost than the main structure, and represent a significant threat for house loss (Leonard & Blanchi 2005). Outbuildings like garages and sheds have more gaps and openings, and are more susceptible to ember ignition. During a fire, they are less likely to be the focus of residents' attention, rendering them more susceptible to ignition. Residents should prepare the surrounds and eliminate ember entry-points prior to a fire event.

Fencing

Post-bushfire surveys have shown that some types of fencing can contribute to the risk to the main residential structure in a similar way to outbuildings (see Fig. 7.6, page 76). In a number of cases in the Canberra fires (2003 survey), fences were responsible for the spread of flames between houses. On the other hand, non-combustible fences can perform as radiant heat barriers. They can reduce the potential for fire attack from the main fire front, low-level ground spread or the burning of an adjacent structure (Leonard et al. 2005).

Recent experiments showed that treated pine fences are readily ignited by ember attack and will most likely burn to completion. Hardwood fences perform better than treated pine fences, and will often self-extinguish as long as there is no adjacent vegetation or combustible groundcover. Open slatted fences, either combustible or non-combustible, provided a negligible barrier to ground-based fire spread and created a partial barrier for radiation. Solid non-combustible fencing systems were very effective in shielding radiation and reducing the potential for ground-based fires to pass (Leonard et al. 2005).

Water tanks

Although water tanks are an important part of water supply for firefighting, in some cases they can present a threat to the main dwelling. Recent experiments have shown that all water tanks perform adequately in maintaining a water supply, provided there is sufficient clearance (of vegetation and other combustible objects) around the tank. Polyethylene tanks are at risk of failure if placed near other fuels, including adjacent polyethylene tanks or structures. Metal tanks have been observed to perform effectively in all situations, even when adjacent to a significant structural fire (Blanchi et al. 2006c).

Influence of human behaviour

Post-bushfire investigations have shown the importance of human behaviour before, during and after the passage of a bushfire front (Leonard et al. 2004). As shown in Table 7.1, a house is more likely to survive with human intervention. Resident and brigade interaction has been identified as the single most significant factor influencing house survival (see Fig. 7.7).

Based on the post-bushfire surveys, there is typically a 3–6 times greater chance of house survival with human intervention soon after a fire front has passed. In most cases, a well-prepared and equipped resident should be able to suppress ignitions by embers before and after the passage of fire front.

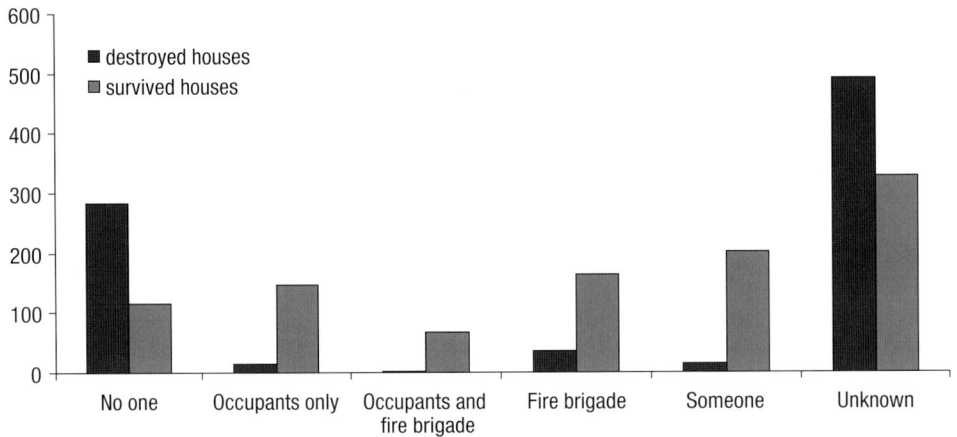

Figure 7.7   Results of firefighting activities during and after bushfire events and house survival (Ash Wednesday 1983, Sydney 1994, Canberra 2003).
Source: Blanchi et al. (2006a).

Table 7.1   Summary of firefighting activities during and after bushfire events and house survival (Ash Wednesday 1983, Sydney 1994, Canberra 2003)

	No one	Someone	Unknown
Houses destroyed (%)	34	8	58
Houses surviving (%)	11	57	32

Fatalities

Fatalities in bushfires have mainly occurred during late evacuation, when people have been caught on the road on foot or in vehicles (Krusel & Petris 1992; AFAC nd). Although taking refuge in a house has consistently proved to be safer than attempting to flee on foot or by car when a fire approaches, there have been fatalities inside houses.

A small number of fatalities have occurred within houses, virtually all of which involved occupants sheltering in a remote location within the structure, away from visual external cues that the fire front had passed. In each case, the house was likely to have remained tenable for some time after the fire front had passed, but the resident had not realised that it was safe to exit and possibly defend the structure or evacuate to safety. People who shelter in houses should be able bodied and have a good understanding of the timeframe in which a fire hazard passes compared to the timeframe when a house can become untenable, and should remain mobile and visually attentive as the fire front passes.

An integrated approach

There are many ways to influence the risk of house and life loss during unplanned bushfires. It is clear that a single initiative such as building code reform is not likely to have a significant effect, but an integrated approach where several initiatives work holistically has great potential. In this way, the responsibility can be effectively shared between the community and government bodies to maximise the level of risk mitigation. The key areas are building code reform, town planning, fire suppression strategies and community education. Policy can play a key role in promoting harmony between these initiatives, ensuring that provisions that are put in place work holistically with all other provisions, and are understood and maintained by the community members they are designed to protect.

Conclusion: outcomes and recommendations

- Bushfire is an inevitable part of living at the rural–urban interface.
- People living in bushfire-prone areas may not have enough warning to plan a safe evacuation and therefore should be familiar with and take steps to mitigate the specific risks their house faces. A house is a safer alternative than late evacuation.
- Able residents can extinguish ignitions before, during and after a fire front passes, significantly increasing the chances of the house surviving.
- Virtually all house losses have occurred during extreme weather conditions, and therefore data on these extreme conditions and their likely frequency in a given region should be used as the basis for preparing for future events.
- Preparedness can be enhanced by an integrated approach, including a balance between house design, choice of materials, landscape design and the maintenance of surroundings. A clear understanding of the interaction between these features and general bushfire knowledge will greatly improve residents' chances of survival.
- House loss follows a weakest-link principle. Effective and efficient approaches to bushfire risk management involve systematically identifying and eliminating the weakest points until there is a suitable level of residual risk.
- Effective risk mitigation measures are best considered in the context of the broader design and performance objectives of the home or urban area (e.g. solar orientation and energy conservation).

Chapter 8

Don't get burnt by the law!

Legal implications of the 'Prepare, stay and defend or leave early' policy[1]

Elsie Loh

Chapter summary

Australian law relating to the powers and immunities of emergency organisations and workers is different in every state and territory. There is no single comprehensive national approach to emergency response in Australia as this area falls within state/territory jurisdiction. This and the continual evolution of law in this area has resulted in much uncertainty among emergency response workers on all levels, from volunteers on the ground to the most senior decision-makers at the agencies. By undertaking a comprehensive analysis of Australian legislation and case law, this chapter looks at the legal implications of the 'Prepare, stay and defend or leave early' policy in each Australian state and territory. It looks at the powers and immunities of members and officers of emergency service organisations (ESOs) under specific emergency services legislation. Despite the complexity and some inconsistency in the approaches taken by states and territories in Australia, members and officers of ESOs are generally given comprehensive levels of immunities, although there are some gaps. ESOs and relevant government bodies may wish to address these gaps, to ensure that their members and officers are protected to a satisfactory level. This overview of ESO powers and protection should also help emergency workers implement the 'Prepare, stay and defend or leave early' policy more confidently.

Introduction

The 'Prepare, stay and defend or leave early' policy ('defend or leave early policy') is a fundamental component of community bushfire safety and widely accepted among Australian fire agencies. All states and territories in Australia have adopted the Australasian Fire Authorities Council (AFAC)'s *Position paper on bushfires and community safety* issued on 28 November 2005. The paper sets out clearly the need for people to prepare their property, then either stay and defend it against a bushfire or leave the area well before the fire is likely to arrive. Though the current Australian legal landscape dealing with emergencies has provided some guidance on this issue, fire agencies still experience practical difficulties in implementing the defend or leave early policy.

The legal aspects relating to fire management may appear complicated due to the changing nature of the common law and the range of relevant fire and emergency service legislation across the state and territory jurisdictions. The apparent complexity of our law often results in many people in the industry feeling confused about what they should or should not do as an ESO or as an ESO member or officer. This chapter will be of interest to members and officers of ESOs who may have to implement the defend or leave early policy, as it outlines their powers and indemnities from liability in evacuating people, and the implications of the defend or leave early policy on these powers and liabilities. This chapter addresses three main issues in relation to the practical implementation of the policy:

- the liabilities of ESO workers exercising their powers of evacuation
- the liabilities of ESO workers in *not* exercising their powers of evacuation
- the liabilities of ESO workers for the advice they give in relation to the defend or leave early policy.

This recognises that ESO workers have a duty to protect life and property. They must, however, do this in accordance with current knowledge and policies. This chapter does not intend to summarise the entire area of emergency law or cover the powers and liabilities of ESOs and their members over crown land (e.g. state forests, national parks, public land). We therefore do not look at the liabilities of land management agencies, such as state/territory Parks and Wildlife Departments that may also have powers to manage fires.

Powers

Specific emergency services legislation gives ESOs broad powers to do whatever is necessary to manage a fire and reduce injury or risk of injury to life and property. These powers may include specific powers to order and direct an evacuation, and even forcefully evacuate people. Generally, two different types of evacuation models are used to describe the evacuation powers of emergency workers. These are the pecuniary interest evacuation model and the mandatory evacuation model (Karanev 2001). Historically, an order to evacuate could be lawfully refused on the basis of a pecuniary interest. A pecuniary interest is a property right that can include goods and chattels. It is based on the principle, dating back to the Middle Ages, that a person who is not a felon or unlikely to act unlawfully can freely enjoy their property rights unencumbered by the state. For example, in *Balfour v Balfour* [1919] 2 KB 571, the court recognised that 'each house is a domain into which the King's writ does not seek to run, and to which his officers do not seek to be admitted' (p. 579). The pecuniary interest evacuation model allows a person with a pecuniary interest in land or in any goods or valuables on the land or in the building to lawfully refuse an order to evacuate. In other contexts the right to refuse an order to evacuate on the basis of pecuniary interest has been overridden. This situation has been described as the mandatory evacuation model, as it is mandatory for a person to obey orders to evacuate despite having a pecuniary interest in the land. It must be pointed out, however, that though a person who has been ordered to evacuate must do so under this model, the decision to evacuate is actually discretionary. There is no blanket legal rule which requires the exercise of this power in every fire emergency. Instead, the power must be exercised in accordance with ESO policies and for the purpose of saving and protecting life and property (which may not necessarily be achieved by evacuating people). This is discussed in more detail later in the chapter.

Victoria is generally described as having a pecuniary interest evacuation model; the other states and territories (New South Wales, South Australia, Tasmania, Queensland, Australian Capital Territory, Western Australia and Northern Territory) have the mandatory evacuation model (Karanev 2001). However, such a dichotomy is simplistic.

Summary of evacuation powers in Australia

Though the pecuniary interest evacuation model has been legislated in Victoria, it is unclear whether it is applicable in all circumstances, such as when an emergency area has been declared. Section 36B(1)(d) and s3 of the *Emergency Management Act 1986* (Vic.) provides that people with a pecuniary interest who have been allowed into their property or are already on the property may have conditions placed on their staying. It is unclear whether these people may be evacuated if they breach the conditions that were imposed on them.

Only in South Australia does the mandatory evacuation model (where people must leave if ordered to do so and force may be used to remove unwilling people) clearly apply in all circumstances, during both 'normal' fire emergencies and declared states of emergency/ disasters. The mandatory evacuation model also applies in most cases in Queensland and in some instances in Tasmania. There are only a couple of instances where the mandatory evacuation model does not apply in Queensland. An authorised rescue officer under s100 of the *Disaster Management Act* (Qld) and a person authorised by the chairperson of the state group or a district disaster co-ordinator under s110 of the *Disaster Management Act* (Qld) may only direct a person to leave. The *Disaster Management Act* (Qld) does not give emergency workers clear powers to forcefully remove people. In Tasmania, police officers have the powers to use force (as reasonably necessary) to remove people who are on land or premises that is burning or threatened by fire. Forced evacuations are also allowed in some jurisdictions when a state of emergency or disaster is declared (such as in New South Wales, Northern Territory and Western Australia). In Tasmania, forced evacuations by fire agencies can only be carried out if authorised by the state controller or the relevant regional controller.

In the other states and territories, the application of the mandatory evacuation model is not entirely clear. In many cases, emergency workers are given the power to direct people to leave, and legislation may even make it an offence to disobey such directions but does not actually give power to the emergency workers to forcefully remove people (e.g. in ACT and Northern Territory). Forced removal of people is often restricted to when their presence is interfering with brigade operations (e.g. in ACT, New South Wales, Northern Territory, Tasmania, Western Australia).

Emergency workers are often given broad powers to do whatever is necessary to protect and save life and property (e.g. in New South Wales, Northern Territory, Tasmania) but it is unclear whether this translates to being able to forcefully remove people from their homes – that would be considered a trespass (such as assault or battery) and at odds with the common law favouring the recognition of people's pecuniary interests. Should force be used in exercising these broad powers, it must be reasonable and not excessive (e.g. knocking someone unconscious in order to remove them would not be acceptable). Care must be taken in exercising evacuation powers in all states and territories, especially those where it is unclear to what extent and in which circumstances the evacuation models apply.

Legal actions

In the event of a fire emergency, general principles which exist in our common law (decisions and reasons for decisions made by judges, based on factual cases) will determine when liabilities are incurred by a rescuer (a member of the fire brigade or emergency service) or by an ESO to a member of the public. Parliament has also enacted legislation that has, in some cases, amended the position of the common law. The extent to which common law has been amended is different in every state and territory, but there is a clear pattern that the common law has been increasingly constrained.

ESOs and their members may be subject to legal action by the public in the implementation of the defend or leave early policy, whether in evacuating, in not evacuating or in giving advice in public in relation to the policy. The circumstances when a person is liable for failure to exercise a power are limited. There are two main types of legal actions – criminal and civil legal actions (torts). A tort is a civil wrong where one party (the plaintiff) alleges another party (the defendant) has done something that has caused harm to the plaintiff and for which they are entitled to compensation. In the context of bushfire emergencies, the torts of assault/battery, trespass and negligence are the most relevant.

Members of an ESO may also be subject to criminal prosecution for crimes including homicide, causing serious injury or assault. However, in order to prove most criminal offences, it must be shown that the person charged had the intention to commit the crime. This will not usually be the case in an emergency response situation. Some crimes require that the defendant only be reckless in relation to the consequences of their actions. Nevertheless, prosecutors must prove to the court that the accused is guilty of the crime 'beyond reasonable doubt', which is a higher threshold than in civil cases. In civil law, the plaintiff only has to show their case on the 'balance of probabilities'.[2] There has never been a criminal case in Australia brought against an ESO officer or member in relation to their role in firefighting and emergency rescue, though it has happened elsewhere. There was a criminal case in the US against an incident controller, who was charged with manslaughter after the death of four US Forest Service firefighters on 10 July 2001 (IAWF 2007).[3]

In an emergency, an ESO has a duty to save and protect lives and property. In such circumstances the defence of necessity may be used to defend a criminal prosecution. It will succeed where the defendant was faced with a choice between complying with the law and allowing great harm to occur, or minimising harm by breaking the law. The defendant must not have done any more than was reasonably necessary in the circumstances, and the harm done must not be disproportionate to the harm avoided. Therefore, though a criminal action against an ESO or a member is possible, it would be unlikely to succeed in Australia.

Torts: negligence

The most common tort action brought in this area is negligence. Negligence is also the action that attracts the most media attention and the tort that most people in the emergency service area are most familiar with. The law of negligence in Australia is changing and is subject to scrutiny from the legislature, judiciary and the community. Any attempt to comprehensively define the circumstances in which emergency services personnel are likely to be found liable in negligence is likely to be quickly outdated. Broadly speaking, a defendant may be found liable in negligence if:

- they owed the plaintiff a duty of care in exercising their powers or performing their duties at an emergency
- they breached that duty by failing to exercise the required standard of care (to take 'reasonable' care)
- the plaintiff suffered loss or damage as a result of the breach of duty.

For a plaintiff to claim that a rescuer or ESO has been negligent and that compensation should be paid, they must first show that the defendant owed them a duty of care. Because ESOs and their personnel are responsible for emergency management, it will often be 'reasonably foreseeable' that a member of the public could be affected by their conduct. This role also means that there will often be a relationship between the ESO or their members and the general public that is sufficiently proximate for a duty of care to arise. ESOs and their members may, therefore, come under a duty of care when exercising their statutory powers, such as those relating to the defend or leave early policy.

In general, Australian law, in keeping with other common law jurisdictions, does not require strangers to assist each other (there is usually no duty to rescue) (see Windeyer J at 66 in *Hargrave v Goldman* (1963) 110 CLR 40).[4] Nevertheless, the lack of a duty to rescue is not a blanket legal principle that covers all situations. There are cases where a duty to rescue may arise, such as where a particular relationship or set of circumstances could lead to the conclusion that the rescuer owes a duty to a person at risk. It is noted, however, that the relationship of emergency worker and a stranger in need of rescue has not been given such recognition. However, a duty of care may still be found despite the absence of such a relationship. There are many factors which the court will consider in determining whether a duty arises. These factors include the extent of control exercised by the rescuer (*Timbs*, 2004), the vulnerability of the person at risk (*Pyrenees*, 1998), the nature of the power held by the rescuer, the degree of risk involved and whether the plaintiff relied on the rescuer's conduct, to name a few (Dunlop 2004).

An example of a situation where the defendant has a large degree of control over the plaintiff or an activity that poses danger to the plaintiff is where an ESO and/or its member directs a home owner to leave their home to an evacuation centre. In such a situation the ESO will have taken on a great degree of control over that person. A duty of care will therefore arise and it would be reasonable to expect that the ESO take appropriate measures to ensure that the evacuation process and centre are safe. There are also risks involved in the ESO deciding not to evacuate (they decide to leave people in their homes to defend their property). These are addressed later in this chapter.

This issue of whether ESOs and members owe a duty of care to individuals is not clearly settled in Australia. This is in contrast to the situation in the UK, where it has been held that ESOs do not owe a duty of care to individuals. It must be noted, however, that Australian courts will in these circumstances tend to set the bar to show negligence higher than usual in recognition that emergencies are special circumstances.

Even if duty of care is found, plaintiffs must still show that the emergency worker has breached that duty by failing to exercise the required standard of care (to take reasonable care). This is determined according to what a reasonable emergency worker would do in the same circumstances. The test has been largely amended in legislation. Therefore, though the court may decide that a rescuer is liable for the harm/damage suffered by an individual, in that the rescuer owes a duty of care and has failed to take reasonable care, the courts and the Australian public in general have always been sympathetic to the cause of emergency workers (e.g. *New South Wales v Brown* [2003] NSWCA 21).

Immunities

In almost all civil cases, volunteers or employees will not face personal financial loss as they will be covered by common law vicarious liability or by its statutory variant. ESOs usually bear the financial cost of their members' actions. In the last few years, there has been increased regulation of liability by statute, and some states now have statutory immunities ensuring that neither the individual nor their organisation is liable. In some legislation, the ESO or member must show that the act or omission was done or made in good faith, to be protected under the immunity provisions. This concept of 'good faith', however, is not defined in legislation and there is only limited judicial guidance on its definition. Courts have described 'good faith' to mean 'without any indirect improper motive' (*Bankstown City Council v Alamdo Holdings Pty Ltd* [2005] HCA 46 at 59). More recently the Federal Court has emphasised the notion of honesty, although this requires more than honest incompetence. A court will generally consider what a person's state of mind actually was, as well as how a reasonable person with the same level of experience and expertise would have conducted themselves in the same circumstances (*Mid Density Developments Pty Ltd v Rockdale Municipal Council* (1993) 116 ALR 460 at 468). It would generally cover acts which are well-meant but unreasonable.

Nevertheless, it is generally accepted that what is required of 'good faith' is less than what is required in common law for liability, which is 'reasonable' (the usual standard). Therefore, volunteers will generally be protected under such protection provisions if they can show their acts were in good faith, even though their acts may have been unreasonable. For example, if a member of a ESO is protecting a house from approaching fires and in good faith – honestly and with the best interests of its owners in mind – decided to pull down a wall in the course of fighting the fire, they will probably be protected even if it turns out that pulling down the wall was unnecessary (a decision that was unreasonable). Protection would not apply, however, if the member accidentally knocked the wall over, as it could not be said in this instance that they had acted in good faith.

The exclusion clauses in Australia can be generally classified into three types:

- those that make no change to the common law (reinforce the existing position)
- those that reinforce the notion of vicarious liability (while extending the principle to volunteers)
- those that make some changes to the common law (Eburn 2005).

Some Queensland and Victorian legislation appear to be re-statements of the current common law position. The relevant sections provide that there is no liability where the act or omission was done 'without negligence' or unless the loss or the injury was caused by 'the negligence or wilful default' of an officer (*Fire and Rescue Service Act 1990* (Qld) s129, *Public Safety Preservation Act 1986* (Qld) s47 and *Emergency Management Act 1986* (Vic.) s37). This is, of course, the current common law position – that the exercise of statutory power which is not negligent cannot attract liability even if damage was caused. On the flip side, where there is negligence (and/or wilful default), there can be liability.

Other legislation appears to reinforce the doctrine of vicarious liability as applying to employees of ESOs. The common law doctrine of vicarious liability provides that an ESO, as the employer, would be liable for acts done by the employee officer, if the member was acting within the scope of their employment or authority. To disprove vicarious liability, the ESO must show that the employee's conduct was so far removed from what was authorised as to be beyond the ESO's control or influence. For example, legislation in South Australia, Victoria and Tasmania provides that a person is not liable for 'an honest act or omission' or an act done or omitted to be done 'in good faith' (*Emergency Management Act 2004* (SA) s 32, *Fire and Emergency Services Act 2005* (SA) s127, *Country Fire Authority Act 1958* (Vic.) s92, *Victoria State Emergency Service Act 2005* (Vic.) s42 and *Emergency Management Act 2006* (Tas.) s58). Similarly, the ACT legislation provides protection for acts done 'honestly and without reckless disregard' (*Emergencies Act 2004* (ACT) s198). An individual's liability is reduced from the test of 'reasonableness' to one of 'good faith' or honesty and 'without recklessness'.

Further, these Acts also state that liability that would otherwise apply to the person should apply to the ESO or the crown (state or territory government). This means that though the ESO or the crown will not be liable for acts done by an employee officer in 'good faith' or 'honestly and without recklessness' (if the employee is not found liable, then under the doctrine of vicarious liability the employer will also not be liable), the ESO or crown would nevertheless still be liable for acts committed by an employee which are 'not reasonable' (the employee officers themselves would also not be liable). This is in accordance with the doctrine of vicarious liability for employees. It would appear that parliament intends these sections to clarify the applicability of the doctrine to employees of ESOs. The doctrine of vicarious liability, however, does not normally apply to volunteers in common law (Ipp Report 2002). As legislation provides the same protection to volunteers as to employees, these provisions effectively extend the doctrine of vicarious liability to volunteers as well. Legislation therefore provides volunteers of ESOs with greater protection than the common law.

Some New South Wales, Western Australia, Victorian and Northern Territory legislation does change the common law significantly by changing the standard of care expected from a duty to take reasonable care to a duty to act in good faith or to act honestly and without reckless disregard.[5] Their effect is to remove the liability of the member concerned even if it can be shown that the conduct was not reasonable. Some of this legislation expressly removes liability from the ESO and the crown if good faith can be shown.[6] Others are silent on whether an action can be brought against the ESO and/or the crown even though the member or officer does not incur liability for acts done in good faith (*Fire and Emergency Act 1996* (NT) s47 and *State Emergency Services Act 1989* (NSW)). Further clarification on this issue by parliament would be desirable.

The policy: prepare, stay and defend or leave early

As discussed earlier, ESO workers must make a choice consistent with policy as to whether the appropriate thing in a particular emergency is to evacuate or not. Before any evacuation powers are used, a choice of whether to evacuate must first be made in light of the accepted 'Prepare, stay and defend or leave early' policy.

Last-minute evacuations

If an ESO exercises its discretion to evacuate in contravention of the defend or leave early policy (such as forcefully evacuating people at the last minute) there is a high likelihood that liability may be established. Such late evacuation would generally not be considered reasonable in light of the defend or leave early policy, which would be what the courts will consider in determining reasonable practice. However, they will also look at the specific circumstances of the case in determining whether evacuation at the last minute was reasonable. Such last-minute evacuation by an ESO may be necessary where the situation would have been riskier for the person to stay than to leave with assistance by the ESO (e.g. if the ESO finds a household in a panicked state, poorly informed and/or ill-equipped, especially if people are on the verge of leaving by car anyway, which would be an extremely dangerous option).

   The decision to leave should ideally be made early in accordance with the policy, at a time when the threat of fire remains distant and steps for leaving homes can be taken without imminent risk or harm. If the ESO exercises its discretion not to evacuate people from their homes in an area of danger and the opportunity to evacuate passes (when danger becomes imminent and the fire front has arrived) then, according to the policy, any question of evacuation is no longer an issue. This is because it is well-accepted now under the policy that last-minute evacuations should not take place.

   In light of this, an ESO or member would have a greater chance of being found liable for undertaking last-minute evacuations than for not evacuating, as this contravenes the defend or leave early policy and may put the evacuees at greater risk than if they had been left in their homes. Members and ESOs acting in accordance with the policy should rest assured that they are acting within the ambit of the law.

Early evacuations

If a decision is made to evacuate early, force should only be used in evacuating if it is clearly allowed in legislation. If the legislation is silent about whether force may be used, ESO workers run the risk of being liable for trespass (e.g. assault or battery) against a person. Otherwise, ESO workers may only direct or order a person to leave where it is an offence for a person to fail to follow such a direction or order. In Victoria, ESO workers cannot evacuate Victorian residents from their homes if they have a pecuniary interest in the land, building or goods in it. An ESO worker may warn and invite a person to leave, but it is not an offence if a Victorian resident decides to stay.

Even if there is clear power to evacuate, an important point is that the decision to evacuate must be considered carefully as evacuating is often an onerous, costly and dangerous task. Further, as forced evacuations involve a degree of deprivation of civil liberties, the power should be used only in situations of great urgency. In such situations, it is extremely difficult to provide the public with the information – such as why an evacuation is necessary and where they are being evacuated to – necessary to obtain informed consent. If informed consent is not obtained, there is potential for ESOs and their personnel to be exposed to actions for trespass against the person. There may also be legal and political fallout regarding the use of reasonable force if a person refuses to leave their home. Kanarev notes correctly that 'it would not be politically acceptable to evacuate a person from their home at gunpoint' (Karanev 2001).

Promoting preparing, staying and defending

The idea of not fleeing in the face of danger is almost counter-intuitive. Nevertheless, it appears to be increasingly popular as more residents are making the decision to prepare, then stay and defend their homes instead of leaving them undefended from oncoming flames. However, this concept is relatively new as a major public policy and the public may still be uncertain about its application. It would therefore not be surprising if ESO workers are asked by particular households whether their expert opinion is that those residents should stay and defend their home or leave early. There are risks involved in conclusively recommending residents to decide to stay and defend. A duty of care is created when an ESO worker holds themselves out as an expert in this area and gives residents a green light to stay and defend, which residents have relied on in making their decision. If harm occurs, injured residents may argue that ESO workers breached their duty by failing to exercise the required standard of care (take reasonable care) by telling them that it was safe to stay and defend when they should not have.

The danger of conclusively recommending residents to stay is that there are many factors involved in a person's ability to effectively defend their home. First, the house may not be prepared sufficiently (even though residents and firefighters may believe it is). This risk may be minimised if the ESO worker undertakes a thorough assessment of the property before giving the advice to stay. Alternatively, residents may not be physically able to defend the property. They may have health problems which the ESO worker may not be medically trained to identify or to determine the extent of a problem's effect. In this situation, it may be argued that it would not be reasonable for the ESO worker to assess whether residents should stay as the ESO worker should know that they do not have the skill to make such a judgment. Further, people may not have the mental fortitude to cope with the stress of the emergency or the actual knowledge (compared to perceived knowledge) of what to do. Arguably, an ESO worker does not have the skill to make psychological assessments and it would therefore not be reasonable for them to give conclusive advice. It is unclear whether it could be argued successfully that such advice was given in good faith and fall under immunity provisions if it can be shown that the ESO worker had attempted to address these issues with the resident. For example, it may be possible to argue that the ESO worker had acted in good faith if they made sufficient inquiries about the physical and mental health of the residents and had attempted to address this issue (e.g. the inquiries and personal observation indicated that the residents appeared fit and able to defend their homes effectively). This would need to be argued in court and will depend on each factual situation.

A clear example of a failure to take reasonable care or act in good faith is where an ESO worker announces to residents in a community that they should all stay and defend their homes without assessing the preparedness of their homes and/or discussing the issues of physical and mental health with individual households. If a resident was disabled or suffered from anxiety attacks, or did not have their house prepared, but still acted upon the ESO worker's advice to stay and as a result was injured or died, there would be a strong case that the ESO worker had acted negligently (unreasonably and not in good faith, as no attempt was made to assess the situation).

It is acknowledged, however, that it may be unrealistic and inappropriate for ESO workers to shy away from giving their opinion of an individual's capability to stay and defend once they have made a positive (and prudent) assessment of the house as well as the physical and mental ability of residents. It may indeed be necessary for the ESO worker to recommend that people leave early if an initial assessment of house, health and/or mental fortitude show that residents are unable to defend effectively. On the other hand, if an ESO worker has a close relationship with a resident and knows that they are not only able-bodied but also have the mental fortitude, knowledge and even experience to cope, it may be appropriate (considering the minimal risk) for the ESO worker to recommend that such a person stay and defend their home. This recognises the fact that ESOs need to weigh the legal risk of making such recommendations against com-munity expectation and the negative consequences (e.g. breakdown in trust and relationship) if ESO workers appear to be unhelpful or shy away from what the community considers their responsibility. In order to minimise risk in making recommendations, it would be prudent for ESO workers to leave the final decision to stay and defend up to residents. They may say that their initial assessment (presuming it was done) shows that the household may look prepared to stay and defend and the house may look defendable, but the decision to stay must be made by the residents because fires are unpredictable and there is always a chance (no matter how small) of death and/or serious injury. Only the residents can decide to take on this risk and they are arguably in a better position to assess their mental and physical ability to defend effectively.

Conclusion

The powers and protection given by legislation differs across state or territory jurisdictions. Powers to force evacuation are not always clearly provided for in legislation. Even if there are clear powers to evacuate, there are many considerations to take into account. For example, in every stage of an evacuation – withdrawal, shelter and return – the ESO personnel involved in the process are likely to assume a duty of care. Thus any stage of the evacuation process may create a claim for negligence. Further, it is contrary to the defend or leave early policy and to published evidence for last-minute evacuations to take place, and acting contrary to well-accepted policies would in most (though not all) circumstances be considered unreasonable. Where evacuations are necessary, they should be carried out early, preferably with the consent of the residents and in a reasonable and competent manner. Care must also be taken when giving advice to people to stay and defend their homes. Though there are risks in giving residents advice to stay and defend their homes because of the numerous uncontrollable variables the ESO worker may not have the expertise to assess (e.g. a person's mental fortitude and physical health), these risks may be minimised.

Where an act (or omission) is unreasonable but done in good faith (or with honesty), in most but not all instances there is legislative protection. There is no doubt that parliament intends to give broad protection to ESOs and their members. However, none of these provisions have actually been brought to court and interpreted. Though the above analysis gives some idea of the immunities for ESOs and their workers, the extent of protection these provisions actually provide (above that given in common law) is yet to be seen.

Acknowledgments

The comprehensive version of this chapter in the form of the CRC Bushfire *Discussion paper: legal considerations of AFAC's stay or go policy* is available on the Bushfire CRC website on www.bushfirecrc.com. I wish to acknowledge Rebecca Monson for her earlier work on the discussion paper as well as the feedback and comments received from various ESOs in relation to the discussion paper and this chapter.

Endnotes

1 This chapter does not constitute any form of legal advice. The author recommends seeking independent legal advice on the issues outlined here. The author will not be held accountable for any decisions made based upon the contents of this publication.
2 Please refer to Luntz H & Hambly D (2005), *Torts: cases and commentary*, 5th edn, for a more in-depth look at the area.
3 As manslaughter charges were only laid on 30 January 2007, the outcome of the case was still unknown at the time of publication.
4 It is a criminal offence in the Northern Territory for a person who is able to provide 'rescue, resuscitation, medical treatment, first aid or succour of any kind to a person urgently in need of it and whose life may be endangered' to 'callously fail to do so'. This offence is punishable by seven years' imprisonment: *Criminal Code Act* (NT) s155.
5 *Fire Brigades Act 1989* (NSW) s78, *State Emergency and Rescue Management Act 1989* (NSW) ss41 and 62, *State Emergency Services Act 1989* (NSW) s25, *Rural Fires Act 1997* (NSW) s128, *Disaster Management Act 2003* (Qld) s144, *Metropolitan Fire Brigades Act 1958* (Vic.) s54A, *Fire and Emergency Services Authority of Western Australia Act 1998* (WA) s37, *Emergency Management Act 2005* (WA) s100, *Disasters Act 1982* (NT) s42.
6 *Fire Brigades Act 1989* (NSW), *Rural Fires Act 1997* (NSW), *State Emergency and Rescue Management Act 1989* (NSW) s41, *Emergency Services Act 1976* (Tas.), *Fire Brigades Act 1942* (WA), *Fire and Emergency Services Authority of Western Australia Act 1998* (WA), *Emergency Management Act 2005* (WA), *Disasters Act 1982* (NT).

RISK PREVENTION AND COMMUNICATION

Painting by Mark Schaller

Chapter 9

Understanding and preventing bushfire arson

Damon Muller and Colleen Bryant

Chapter summary

While bushfire is a constant fact of life in Australian society, little is known about those who deliberately light bushfires. The purpose of the bushfire arson project being conducted by the Australian Institute of Criminology is to understand bushfire arson, including those who light bushfires and why they do so.

This chapter aims to present some of the project's initial findings, within the context of what exactly arson is and how common it is, based on an analysis of data provided by Australian fire agencies. The motivations behind arsonists are discussed, including the difference between bushfire and non-bushfire arson, and the differing motivations of children and adults. It also discusses the ways in which we deal with bushfire arsonists, including options for the criminal justice system, and findings from a survey of arson intervention programs in Australia, particularly for juvenile arsonists.

Given that up to 60% of bushfires in Australia may be deliberately lit, an understanding of bushfire arson and responses to arson may help fire services, the criminal justice system and communities appropriately identify, respond to and prevent deliberately lit bushfires.

Introduction

Bushfire is an ever-present reality in the Australian landscape. Many of Australia's ecosystems have evolved not only to withstand the effects of fire but have exploited fire for their own survival. Perhaps fuelled by this knowledge, and the fact that natural fire causes have played a major role in the large-scale disasters involving significant loss of life and property, it is easy to be under a misapprehension that fires in the Australian landscape are essentially a natural phenomenon. Belying this assumption is the fact that only 5–6% of the 45 000–60 000 fires attended by fire agencies every year across Australia (excluding fires in the tropical savannas) are attributed to natural causes (Bryant 2007; Productivity Commission 2006). Even less are the result of lightning. That is, 95% of all fires in the Australian landscape are the result of human actions, with few resulting from the planned use of fires. Approximately half of all fires where causal attributions were made (approximately 60% in total, but variable between jurisdictions) resulted from incendiary activity or were suspicious in origin (Bryant 2007). Such figures do not include fires attributed to children or the majority of fires attributed to smoking-related materials.

Although the majority of deliberately lit fires are small in size (less than several hectares; Bryant 2007) and typically do not result in the loss of human life or property, the costs to society and the environment reach far beyond the economic statistics. Even if unintentional, there is an ever-present risk that, under adverse or unpredictable conditions, fires have the potential to escape and cause great damage. In reality, substantial economic costs are incurred by fire agencies and thus the governments and taxpayers who fund them. In many regional areas many of those costs are borne directly by the volunteers and their local community. Further firesetting during existing bushfire events diverts resources and places unnecessary strain on already-stretched agencies and individuals, increasing the risk that other fires will spread. Although prosecution of offenders is generally welcomed it is not an inexpensive exercise, affecting not only police, judicial and correctional services but also the fire and land management agencies who provide the specialised fire investigation resources and briefs that enable prosecution to proceed.

A high proportion of all bushfires in Australia do not take place in the 'bush' per se, but in local parks and open spaces or remnant fragments of vegetation within or near cities, and along the interface between urban and rural/regional landscapes. There is a clear relationship between high populations and increased probabilities of deliberate vegetation fire in Australia. Typically, 40–50% of all fires occur in vegetation within or surrounding major metropolitan centres, and a high proportion of the remainder are associated with major or smaller urban centres (Bryant 2007). There is necessarily a flow-on effect to the surrounding vegetation, and that damage can be substantial.

Although many Australian ecosystems have adapted to fire, the higher frequency and more haphazard timing of fires in the urban and semi-urban environment represents a marked change in the natural fire regime (Cheney 1995). Few ecosystems respond positively to the high frequency of fires in many urban or semi-rural environments. Not surprisingly, the losses may be even greater in more delicate ecosystems, which are unaccustomed to and therefore evolutionarily ill-equipped to deal with fire. Severe events may not only result in loss of biodiversity, but destroy habitats and facilitate the invasion of weeds. They may also disrupt hazard reduction programs conducted by land management agencies; programs that are specifically designed to protect vulnerable species or communities in the face of large-scale natural fires.

What is bushfire arson?

Strictly speaking, arson is a criminal act and can only be so defined by a successful action in the criminal justice system. Due to the inherent ambiguity in determining the cause of some fires and the difficulty in successfully prosecuting a crime for which the offender essentially needs to be caught red-handed, many fires that are suspected of being deliberately lit will never be confirmed as arson. Many people who illegally light such fires will never be charged with, or convicted of, an arson-related offence.

Traditionally, the term arson implies the concept of malicious intent. As fire remains central to human interactions with their environment, through land management practices, recreation (barbecues and camping) or other activities, the issues of intent and accountability are extremely complex in both determination and prosecution. There is a spectrum from what may be regarded as accidental through carelessness and recklessness to malicious intent. This is true of both bushfire and other forms of arson, such as structural arson (setting fire to a building or car), but bushfire arson involves a unique set of problems. These primarily relate to the large amount of damage that a bushfire can cause and the potential for the fire to get out of control, regardless of the intention of the person who lights it.

Some states have circumvented the issue of malicious intent in bushfire arson to a certain extent by adopting key elements of the Model Criminal Code Officers Committee's (MCCOC) model code which relate to bushfire arson (MCCOC 2001). Bushfire arson in the code differs from other types of arson in that it focuses on the potential for catastrophic destruction from bushfires, rather than the actual damage encompassed in other arson offences. It does not need to establish an intent to endanger life, merely proof that an offender was reckless as to the possible consequences of their actions (MCCOC 2001: 53). However, this clause is by no means definitive, and there are issues with the level of children's awareness about the potential dangers of fire.

An in-depth discussion of the definitional issues surrounding bushfire arson is beyond the scope of this chapter (see Willis 2004 for more details). In the absence of a broader and more convenient definition, 'bushfire' refers to any bush, grass, forest or other vegetation fire, regardless of the size of the area burnt, and 'bushfire arson' is any such fire where there was an intent to, or a disregard for potential to, cause damage, regardless of whether the person is identified or prosecuted. Those who light such fires, irrespective of whether they are charged, prosecuted or criminally liable, are referred to as 'arsonists'.

Figure 9.1   Public outreach sign on the Hume Highway, just on the outskirts of Melbourne near Wallan, Victoria. © Bushfire CRC.

Why do people light bushfires?

The quest to understand why people light destructive fires has inspired considerable research, although little has been directed toward understanding the context for lighting such fires in the natural environment, in Australia or internationally. The profiles of arsonists who commit structural arson are not necessarily relevant, or are of minor significance in relation to bushfire arson.

Adult arsonists

There are a variety of reasons that people might deliberately light bushfires, and thus a variety of profiles of arsonists. Although there is overlap, the dominant motives for bushfire arson tend to differ from those for other forms of arson. For example, structural (non-bushfire) arson is often committed for revenge or for financial reward, especially through insurance and other fraud. Although such motives do occur they are less likely to motivate bushfire arson, and even where present have a unique flavour. For example, in cases of revenge, anger is not necessarily directed at a specific individual but at society at large (Shea 2002). Financial rewards may also be indirect. The financial benefits derive from the additional overtime hours or because the fire opens up land, use of which existing regulations had prevented. In some cases the damage is inadvertent. For example, in torching an abandoned car in a state forest the intention might be to conceal a crime, not to cause a bushfire, but the impacts on vegetation can be the same.

People who light bushfires generally do so because it fulfils a psychological need (Shea 2002). This is not to imply that all bushfire arsonists feel an irresistible impulse to light fires, and research suggests that true pyromania is rare or nonexistent (Doley 2003). Rather, by lighting fires people seek a psychological benefit, something they perceive as lacking in themselves. For example, some people light bushfires to boost their self-esteem, as an exercise in power, feeling a sense of control as the fire services race to contain the damage. Others want to bring attention to themselves or to present as a hero by alerting the population and authorities to the fire. Some bushfires are lit to generate excitement, to seek stimulation from the sights and sounds of the fire and those who arrive to fight it, or to relieve boredom. In some cases, poor impulse control or a low capacity for abstract thinking and a lack of appreciation of the outcomes of actions are associated with mental disorders; they may lead to firesetting and other antisocial behaviour (Shea 2002). Although there are a small number of instances of people with a mental illness or intellectual disability lighting bushfires, such disorders do not cause people to light fires. It has also been suggested that lighting bushfires may take the form of politically motivated violence (Baird 2006), although no such cases have been confirmed.

A serial arsonist, who lights a series of bushfires, may be motivated for any of these reasons. Generally, if lighting the bushfire has the desired effect (recognition, excitement etc.) then the firesetting behaviour is likely to continue, as the behaviour is reinforced and the underlying psychological inadequacy remains. The proportion of fires started by serial arsonists may be comparatively high. A study of NSW Rural Fire Service investigation data indicated that approximately 13% of investigated fires were probably the result of serial arson (AIC 2005b), although this figure is not necessarily representative as it concentrated on a subset of fires where an investigation was warranted. Nevertheless, identification and investigation of serial arson can be difficult in the absence of distinguishing characteristics. Fire investigators and police must therefore be alert for patterns when investigating fires, which might indicate the operation of a serial arsonist.

Despite being a topic of interest for a number of years, bushfire arson is still not well understood and insights from research into structural arsonists have only limited relevance. What is known about bushfire arsonists is drawn from those who have been caught – only a small proportion of all arsonists, due to the difficulties of identifying and prosecuting bushfire arson. It is something of a truism that only the 'dumb' offenders are caught – it is possible that the most successful bushfire arsonists will never be identified or understood.

One group of adult arsonists which has been the subject of much attention involves people within the ranks of the fire services. Like a paedophile will attempt to secure a job that involves contact with children, it is to be expected that those with an interest in lighting fires will attempt to join the fire services. Such arsonists throw suspicion on all firefighters and undermine public confidence in the fire services. This is especially pertinent in rural volunteer fire services, where pools of potential volunteers are restricted and resources for vetting them are equally limited. Unfortunately, there is no easy way to identify potential arsonists, so vetting often relies on local knowledge and judgment. While criminal-record checks for volunteers have been suggested in some jurisdictions, the low detection and prosecution rates for arson offenders means that the chance of any given arsonist having a formal criminal record is low. Motives for firefighter arsonists may include seeking recognition or fame for their heroics, or a desire for excitement. They may be as prosaic as a desire to create extra income through overtime pay or to impress others enough to gain full-time employment within fire services. Contrary to media portrayals, potential arsonists make up only a tiny proportion of fire service personnel, but fire services must remain vigilant against the possibility. Close contact with, and observation by, fire personnel increases the likelihood that such arsonists will be suspected and investigated.

Juvenile arsonists

A large proportion of all deliberately lit fires, estimated to be as much as 60–75% in the US (Stanley 2002), are lit by children. The reasons why children light fires are generally considered to be distinct from adult arsonists, as many fires deliberately lit by children may not be with malicious intent. However, with the transition from adolescence into adulthood the potential for overlap between the two may increase.

The issue of why children light fires is complex, potentially linked with the psychological development of the child. Fireplay among children is quite common, although there is some disagreement as to exactly how common it is. Kafry (1990) found that most young boys had an interest in fire, and 45% engaged in fireplay. In a recent Australian study, Dadds and Fraser (2006) found that 42.1% of 7–9-year-old boys showed fire interest, 13.2% had engaged in match play and 5% had a history of fireplay, with lower rates among girls and younger boys. While an interest in fire is not in itself unusual or dangerous, children often have a poor understanding of the real danger, an inaccurate perception of their capacity to extinguish a fire and may be unaware of the potential losses from bushfires, including the dangers posed to other people, themselves, property and wildlife.

It is generally assumed that tendency for setting fires decreases with psychological changes in the maturation process, including cognitive growth, positive social learning experiences, the development of moral reasoning, empathic concern for others, a capacity for means-end thinking and a greater appreciation for negative consequences of antisocial behaviour (Epps & Hollin 2000). Effectively, there may be a progression in the style of activities in which children are likely to join, based on age, from fireplay (less than 7 years of age) to curiosity firesetting (8–12 years of age), with some young persons going on to light at least one fire with the intention to cause damage. A fraction of these will continue to light fires with the intention of causing damage. The National Association of State Fire Marshals (2001) indicates that children who light fires generally fall into two categories: those who are curious about fire and whose firesetting gets out of control, and those who intentionally set fires. Although delinquent behaviours are sometimes adolescence-limited, with children effectively growing out of it as they mature, in others it is life-course persistent (e.g. Scholte 1999).

Lewis (1999) suggests that the most common reason for deliberate firesetting by children is boredom. Children may derive benefit from the experience of watching a fire, whether excitement or gratification. However, Swaffer and Hollin (1995) found that only one of 17 juveniles interviewed was fascinated by fire. Additional benefits may derive from the social context of firesetting, particularly if that occurs as part of a group or with a friend, or from the consequences of firesetting, being either a means to achieve a goal or an expression of intense personal feelings such as rage and anger (Epps & Hollin 2000). From a humanistic perspective, each represents a failure to provide children with constructive means to fulfil their potential.

Research indicates that children are not equally likely to deliberately set fires. Firesetting behaviour has been associated with parental stress, hyperactivity and cruelty to animals, along with other characteristics known to be predictors of antisocial behaviour (Dadds & Fraser 2006). Profiles of juvenile arsonists such as those discussed by Drabsch (2003), including characteristics such as lower socio-economic status, single-parent families, poor self-esteem and a history of abuse, are not substantially different from those of any other young person exhibiting antisocial behaviours. Problematic juvenile firesetters are unlikely to be a distinct group of offenders; rather, firesetting is one aspect of a general pattern of antisocial behaviour engaged in by a small number of young people. Firesetting may continue if the behaviour is reinforced or the psychological need remains unmet.

Punishing and treating arsonists

Arson is considered to be a serious indictable offence in every Australian jurisdiction and carries heavy penalties, including life in prison. In addition, most jurisdictions have legislation specifically targeting the lighting of bushfires. There are also a number of summary offences (crimes which can be tried before a magistrates/local court, without a jury) associated with bushfires, such as lighting a fire on a day of total fire ban, or dropping a lighted cigarette. Due to the lack of distinction between bushfire and other arson in criminal justice statistics, it is difficult to determine the nature of penalties for deliberately lighting a bushfire.

A number of explicit and implicit principles underlie the sentencing of criminals, including rehabilitation, denunciation, deterrence (general and specific) and retribution. As bushfire arson has such a strong emotional impact on those affected by it, media commentators often focus on the retributive aspect of sentencing, but it is important to realise that this is not the only principle under which courts operate. Denunciation and general deterrence (deterring others from attempting the same crimes) also argue for meaningful custodial sentences, especially when those sentences are highlighted by the media.

As discussed earlier, there are a number of reasons why someone might deliberately cause a bushfire. If a bushfire is lit in order to secure fame or recognition for the offender, for some form of financial reward or because of negligence, then specific deterrence – stopping that individual from committing a crime again – may play an important part in sentencing. A few moments of fame or excitement are unlikely to be worth another gaol term or hefty fine.

While certain sections of the criminal justice system and popular opinion are wary of the idea of rehabilitation, there is a strong argument for some attempt to be made with bushfire arsonists. If a person is lighting a fire because it fulfils a psychological need, or if they have some form of intellectual disability which prevents them fully appreciating the consequences of their actions, they are unlikely to cease lighting fires when released without some form of intervention. Our inquiries were unable to determine whether formal arson intervention programs were running in adult prisons in Australia, but forensic mental health services may offer such programs as outpatient treatments. For example, Forensicare in Melbourne has treated arsonists. The underlying causes of arson are poorly understood, however, so firesetting is treated the same as most other problematic behaviours. We are unaware of any evaluations conducted on adult arson intervention programs in Australia.

When dealing with a juvenile (a child aged less than 18 years, or less than 17 years in Queensland), there are additional considerations. Juvenile justice legislation in most Australian jurisdictions is premised on the assumption that imprisonment should be used as a last resort when dealing with children – other options or diversions, such as cautioning or family group conferencing, should be used in preference. Further, children under 10 years of age cannot be held criminally responsible for their actions (AIC 2005a).

The AIC recently completed a survey of arson intervention programs in Australia (Muller & Stebbins 2007). Although most are not formally aligned with the criminal justice system, each Australian jurisdiction operates programs targeting juvenile firesetters (Table 9.1). The programs are essentially based on the two approaches of fire education and behaviour modification, although most contain elements of both. Fire education seeks to inform children about the effects of fire and its safe use. Behaviour modification generally utilises some form of cognitive behavioural therapy, which recognises the link between thoughts and attitudes (cognitions) and behaviour, and uses that link to help change the behaviour.

The programs utilise specially trained firefighters as program facilitators, are often based in the child's home and encourage or require parental participation. The programs also maintain links with mental health services – they often have mental health professionals on staff – and the criminal justice system. Many accept referrals from the youth/children's courts or from family group conferences, although the Queensland-based Juvenile Arson Offenders Program is the only Australian program of which we are aware that exclusively takes referrals from the criminal justice system. At the time of writing, none of the programs had received a formal independent evaluation of their effectiveness, but such evaluations are underway in Victoria and Queensland.

Table 9.1   Juvenile arson intervention programs in Australia

Program name	Jurisdiction	Operating agency
Juvenile Fire Awareness and Intervention Program (JFAIP)	Australian Capital Territory	Australian Capital Territory Fire Brigade
Intervention and Fire Awareness Program (IFAP)	New South Wales	New South Wales Fire Brigades
Juvenile Fire Awareness and Intervention Program (JFAIP)	Northern Territory	Northern Territory Fire and Rescue Service
Fight Fire Fascination (FFF)	Queensland	Queensland Fire and Rescue Service
Juvenile Arson Offenders Program (JAOP)	Queensland	Queensland Fire and Rescue Service
Juvenile Firelighters Intervention Program (JFLIP)	South Australia	South Australia Metropolitan Fire Service
Juvenile Fire Lighter Intervention Program (JFLIP)	Tasmania	Tasmania Fire Service
Juvenile Fire Awareness and Intervention Program (JFAIP)	Victoria	Metropolitan Fire Brigade/ Country Fire Authority
Juvenile and Family Fire Awareness (JAFFA)	Western Australia	Fire and Emergency Services Authority of WA

The correct criminal justice response to bushfire arson, especially when committed by children, is a difficult issue. The media reflect the views of much of the public when they call for harsh punishments for people whose actions have caused huge amounts of financial and ecological damage, but those calls are often inconsistent with the principles on which the criminal justice system operates. Although often perceived as a soft option, a treatment program which may eliminate future arson attempts is unarguably a better outcome for both the offender and the community. Striking a balance between sentencing principles such as denunciation and rehabilitation will continue to be a challenge for the criminal justice system.

Conclusion

Many of the bushfires that plague Australia every season, destroying land, property and lives, are deliberately lit. Identifying who is responsible for a deliberately lit bushfire, and in some cases whether a bushfire was in fact suspicious, is a challenge for police and fire service investigators. Even if an offender is identified, collecting sufficient evidence to prosecute them successfully – short of an eyewitness observing them set a fire – is extremely difficult. As a result, bushfire arson has a low clearance rate, and what we know about arson and arson offenders is the result of a skewed sample.

Although there are harsh criminal penalties for bushfire arson, there is little in the way of treatment for adult arsonists in Australia. A much wider variety of treatment approaches is available throughout Australia for young people who engage in dangerous fireplay, and such programs may prevent a young person with an unhealthy interest in fire from becoming an adult arsonist.

While steps have been taken to understand the prevalence of arson in bushfires generally in Australia, little is known about people who deliberately light bushfires, or their reasons for doing so. We are constantly asked by the media to provide some insight into the mind of the arsonist – a fair question, given that people who have been victims of bushfire arson want to know why they were so targeted. At present, we can only respond to such queries with generalities, particularly when it comes to adult serial arsonists.

The next step in research into arson must be to examine the arsonists. Due to the low detection rate of bushfire arson, and the lack of insight that most offenders have into their own behaviour, finding and interviewing arsonists is challenging. Less direct methods of investigation, such as examining crime statistics, may be a profitable avenue of inquiry.

Although challenging, understanding arsonists is essential for a number of desired outcomes. Knowing the characteristics common to arsonists will assist police and fire investigators in investigating arson, and will help identify arsonists during intake procedures for fire service personnel. It will also assist in determining the appropriate treatment for arsonists.

Chapter 10

The media and fire services
Dealing with conflicting agendas[1]

Erez Cohen, Peter Hughes and Peter B White

> The discourse around fire is saturated with superlatives, with words such as 'exceptional', 'unprecedented', 'extraordinary' and so on. Such hyperbole may give some immediate comfort to those who have just suffered trauma and major losses. But in the long run, it only serves to reinforce ignorance and losses in the inevitable future event (Campbell 2003: 246).

Chapter summary

This chapter seeks to understand the complex relationships between media news organisations and fire and land management agencies in the provision of bushfire warnings and information.

Using a combination of interviews (with journalists and agency staff) and observation (in CFA headquarters and a media unit) this research project identified that relationships between the media and fire-related organisations are now more complex and productive than earlier literature on this subject suggested.

We recommend that these productive relationships be strengthened by more regular monitoring of the media by fire agencies, by joint activities involving journalists and fire agencies, the development of a single media unit to represent all agencies during emergencies, and by developing within the Department of Sustainability and Environment, the land management agency, a specific, clearly branded and separate fire division.

Introduction

The media are identified in the research literature and by Australian fire and land management agencies as high-priority sites where information and misinformation about bushfires is disseminated. Different media organisations have their own interests and agendas in reporting such events. These are often quite different from the interests and agendas of fire and land management agencies. Nor should agencies be regarded as all sharing the same interests, as each agency operates in a particular social, political and cultural context. Land management agencies in particular have complex responsibilities and relationships with the public. The complexity of these responsibilities and relationships can complicate the way that their activities are perceived by the public and reported in the media. This chapter, based on interviews with fire and land management agency staff and with journalists who have reported on fires, identifies the complex relationships fire and land management agencies have with the media and proposes strategies for acknowledging and addressing these different needs. It deals with journalists' perceptions of fire and land management agencies and how those agencies are presented in the media. Although an examination of 'perceptions' might be seen as less than substantial, our view is that identifying and understanding the perceptions of fire and land management agencies held by journalists and the public is important when those agencies are designing and implementing their public communication strategies. After identifying the perceptions we suggest ways in which agencies might address the issues these perceptions raise.

Thinking about bushfires and the media

Andrew Campbell calls for a new understanding of bushfires in Australia. Fire should not be perceived as a 'terrifying aberration, an ineluctable, unpredictable Act of God'; instead Australians should start to see it as 'an inherently Australian phenomenon that goes with the territory' (2003: 244). The language of warfare, destruction and terror should be replaced with the language of knowledge and fire management (Campbell 2003: 244). The same idea is evident in the 2004 report of the Council of Australian Governments (COAG) *National inquiry on bushfire mitigation and management* (Ellis et al. 2004) The logic of this argument is clear. If you teach Australians to 'live with fire' and see it as a part of their natural environment, they are more likely to know how to prepare and what to do when the inevitable fire strikes. Such an understanding will certainly assist fire agencies' aims of achieving better fire management and fire control strategies that may help prevent or manage the risk from large-scale bushfires. In the COAG inquiry's vision for 2020 Australia.

> All Australians understand, accept and respect bushfires and know that they will continue to occur. We have drawn on indigenous, local and scientific knowledge in learning to live with bushfires. Communities understand that the risk, and the responsibility for bushfire mitigation and management, is shared by individuals, landholders, communities, fire and land management agencies, researchers, and governments (Ellis et al. 2004: ix).

The position of the media in promoting different images and understanding (or misunderstandings) of bushfires and their role as a means for delivering safety messages suggest that they cannot be ignored in realising this vision. However, any examination of the media needs to recognise that different social groups (including fire and land management agencies) will have different interests, and different perceptions of the roles and performance of the media.

Figure 10.1   Negative perceptions in the media about the Canberra fires, January 2003.

Mass media images and reports about bushfires are often identified as examples of the problem with the public discourse about bushfires. For example, Campbell argues that 'pervasive formulaic media images and clichés compound such ignorance' (2003: 246) whereby the public cannot really understand the difference between a specific hazard and risk, and thus lack respect for fire. Yet, despite such insights and criticism of media coverage of fires there are often no suggestions on how mass media itself should be targeted. Criticism of the media fails to provide an analysis of why and how different media organisations 'frame' particular events. As Klinenberg argued in analysing the media treatment of a devastating heat wave in Chicago:

> Scholars of social problems typically conduct surveys of news coverage to assess how journalists represent issues or events for their audiences, but they rarely analyse the conditions in which reporters and editors produce their accounts. One consequence of this research strategy is that social scientists do a more convincing job of *showing that* news organizations *distort* their coverage or omit key issues from their stories than of *illustrating how* journalists come up with such representations (2002: 191, original emphasis).

'Media myth' or media misconceptions about bushfires, critics argue, fuel the current emotive (or non-scientific) bushfire public discourse. 'The media' have been repeatedly accused by emergency services and researchers of generating fears and myths about bushfires and other natural disasters (Goltz 1984; Blong 1985; Quarantelli 1989; Marshall 1994; Country Fire Authority of Victoria 2000). 'Exploding houses', 'fire storms', 'fire balls' and 'panic evacuations' are some of the regularly repeated media myths that fire services have been working hard (and some may argue unsuccessfully) to dispel.

Fire agencies fear that irresponsible sensationalist reporting may cause people to try to flee a fire when it is far too late to do so. There is a real concern that media coverage, such as talkback radio shows, may run counter to the goals of emergency organisations both in relation to fire preparations and the actions of the public during fire incidents. On the other hand, lack of reporting or media interest in fires may result in the public being unprepared for fire events or unaware of what actions to take when fire strikes. As a consequence the role of the mass media becomes very important for fire agencies, both as a way of delivering information about bushfires and as a sphere where public perception of fire agencies and bushfires is debated in a highly politicised context.

In this context it is important to make some distinctions between agencies. In this chapter we refer to two Victorian agencies in particular: the Country Fire Authority (CFA), whose responsibility is primarily fire control and mitigation; and the Department of Sustainability and Environment (DSE). As a land management agency the DSE has more complex roles in the community. These include the management of coasts and marine environments, forestry, land management, fire, plants and animals, water, property titles and maps, recreation and tourism. DSE has both planning and enforcement responsibilties in areas which will affect rural and, to a lesser extent, urban residents in a range of ways. As a government department is may be viewed by the community as a more 'political' agency than the CFA, as DSE is required to implement government policies which may be unpopular in sections of the Victorian community. While the 'brand' of the CFA is reasonably clear, the 'brand' of DSE is much more diffused and problematic for some of the people it is seeking to reach. It is likely that the ongoing public perceptions of the agencies and the media strategies used by the two agencies will differ in some respects.

Managing media images: the process

Along with legal scrutiny and litigation, growing community expectations and criticism of fire and land management agencies aired in the media mean that most emergency organisations now consider it necessary to actively engage with the media. Most have public relations and media departments that are both reactive and proactive in promoting the organisation's 'brand name' and in managing community and public expectations via the media. While there are obvious variations in the different agencies' policies for dealing with the media, these aspects of the relationship with the media are well-established.

The fire and land management agencies' view of the media is far from unified. Different states and territories in Australia face very different levels of fire risk, different geographical and social settings and different policies and resources that generate very different relations with the media. This is in addition to internal and changing approaches in particular organisations.

In our work with fire and land management agencies in Victoria, interviewees articulated their organisation's approach to the media as the 'old approach' versus newer and more collaborative approaches to the media. The 'old approach' was defined as suspicion of the media. This culture was reported to be disappearing from the agencies and the new approach was being implemented in two ways.

The first was that of PR-driven professionals whose view is that a careful media strategy is the best way to promote and protect the organisation's public profile. Accordingly, a good PR strategy needs to promote the organisation's goals in any given season in the prevailing social and political environment. During the summer or fire season, the PR department's activities will be different from its activities during the winter. It is also clear that there are important distinctions between the proactive and reactive aspects of such work. Such a view entails a centralised corporate media management plan in order to protect the organisation's brand name and deal more effectively with the challenging political environment and demanding culture of community expectations.

The other approach to the media was presented as an information-flow model placing the emphasis on rapid dissemination of information to the public during incidents. Such an approach views the media as an effective tool for delivering specific warnings and facts (rather than 'spin' or institutionally self-serving messages) about the fires. This model argues that a decentralised media management system is the most effective way to deliver rapid and accurate information to the public during incidents.

Fire and land management agencies in Victoria: a case study

As we have suggested, the management of each agency's relations to the media partly reflects its organisational structure. In Victoria the difference between the two agencies that deal with bushfires – the DSE and the CFA – is interesting in this regard. Journalists we interviewed had a clear perception that the agencies had different media strategies. This case study discusses the perceptions of those journalists, as their professional behaviour is influenced by their perceptions. The DSE, a government department engaged in land management, can centralise and compartmentalise its public communications, but the nature of an organisation like the CFA, which is supported by large numbers of volunteers, makes such a centralised media management practice almost impossible to maintain. This point was raised by a journalist who is also a CFA firefighter volunteer.

> The CFA in the past decade placed an enormous importance on dealing with the media and there has been acceptance within the organisation that the media is one of their tools, particularly in communicating risk to communities. Head office has a media relations section and strategies that are designed to achieve their aims. But when you go down the lower food chain of the local brigade level you'll see that some fire captains and firefighters have embraced that idea and others far less. There is no way to enforce it. Most brigades would have a policy internally about who talks to the media, but the nature of

volunteers mean that you cannot really control them or discipline them (journalist and CFA volunteer).

The organisational priorities and practices regarding communication that best serve the interests of fire agencies may be the very features that journalists consider to be obstacles. Some journalists argued that it is very difficult to work collaboratively with the DSE because, unlike the CFA, the department is overly protective and does not open itself up to media scrutiny.

> The CFA seems to work with the media a lot better then the DSE. The DSE has this sort of governmental bureaucratic approach ... The CFA is much more pragmatic, it is living in the real world. They've got a disaster coming, they know that the media are doing what media does, which is go to where the trouble is to report on it. ... They understand the reality of the media a lot better (*The Age* journalist).

Predictably, any suggestion of fire agencies using 'spin' was viewed negatively by journalists.

> We live in a world of public relations and media people ... their job is to assist the media but what we find a lot of the time is that their job is to actually stop us from doing what we need to do. If I am in a fire my job is to go out and report that fire as accurately as I can and I will not let anything stop me from doing that (*The Age* journalist).

Another journalist from *The Age* argued that DSE's attempts to promote or protect its organisation put it at odds with the media.

> The organisation's public image is one of the main issues for the DSE and it is partly their own fault. If they were better at managing the messages that went out about things like the recent Wilson's Promontory fire they wouldn't have such a problem. Instead, they take a defensive position and therefore they are on the back foot ... It's not that long ago that the CFA had a similar public image problem (*The Age* journalist).

The controversy surrounding the Wilson's Promontory fire of 1 April 2005 exemplified these issues. The fire had burned approximately 6200 hectares by 12 April. The fire had been lit as part of a planned fuel reduction burn. Yet, once the fire had 'escaped', it was presented and interpreted in the popular media as a disaster. The DSE was directly blamed by the media for destroying the park.

> The Prom fire was the worst situation I have been involved in. It had everything for a bad media story. It started through a fuel reduction burn that the DSE had been conducting eleven days before. It involved the evacuation of 600 people. First and foremost, the Prom is Victoria's most loved national park, it is a favourite playground in Victoria. Number two it was school holidays. Number three, it closed the park down for the whole school holiday. Four, it involved evacuation. Five, it started through one of our burns. So the media have jumped all over us. Reporting the fire was one side of things but the other side was putting their boots into us, looking for blame and that sort of thing (DSE staff member).

Despite this damaging media coverage a journalist who had previously had very negative experiences with DSE found that its personnel were very accessible after the Prom incident.

I have just come out of the Wilson's Prom fire and their attitude to the media is a complete turnaround. They couldn't have done enough for us. You could feel that there was a shift in the way that they were looking at the media. They weren't looking at us as if we were their enemy. It was like 'OK, these people are here, they have their job to do, just as we have our job, and instead of treating them like enemies we should work with them'. And they did just that over a four-day period. They choppered us into the fire zone and flew us out and they couldn't have done enough for us. So I thought this was an amazing 180° turn from what I had experienced in 2003 and hopefully that relationship will continue (*The Age* journalist).

Interestingly, a report by the Emergency Services Commissioner, Bruce Esplin, into the origins and handling of the Wilson's Promontory bushfires indicated that the DSE did not follow its own policies for prescribed burns when it started the fire.[2] The response from another Age journalist to this report reflected a perception that the DSE had been manipulating the media during the fire event but was 'caught out doing it', suggesting that the negative and critical media treatment of the DSE was warranted. According to this journalist, DSE was manipulating the media on one front, facilitating media access to the fire so that journalists and photographers could generate 'great pictures' but had to guess the 'real story'.[3]

I think they had learned how to manipulate us more subtly. They now seem to know that we have tight deadlines and that it is difficult to check the veracity of what they say in just a few hours … this is an interesting place they have come to (*The Age* journalist).

This rather cynical view reveals the difficulties faced by multi-functional land management agencies such as DSE. On the one hand DSE was praised for providing ready access to information during the event, but on the other hand the information was seen as 'spin' that served the interests of the organisation. The criticisms expressed by this journalist also ignore the fact that there are different and changing approaches to the media in different fire and land management agencies. It also ignores the fact that there may well be differing policies within an agency, with decisions made at a local level perhaps being perceived by journalists as representative of DSE as a whole. The complex balancing act required to manage the media effectively is described well by a CFA staff member.

I do not think the media is our friend but I do not think the media is our enemy as well. The media also have a number of responsibilities to public information and to various lines of inquiries which this organisation [CFA] would not like to follow. There is an ethical responsibility on journalists to pursue the truth even when it is uncomfortable to organisations like this. I think that means that it is quite important that the relationship between emergency services and the media is never one of friendships. Because they are going to have to really give us a good poke on a couple of issues … They can be a colleague, they can be supportive, they can understand our issue but they have to challenge us, that's their job. So when people say a lot of the time that the media is our friend it gives a false impression that you can bring the media to a point where they will do what we want them to do. The day that happens, we are in a lot of trouble as a democracy. It is not the role of the media to do what emergency services want them to do.

> Does that mean the media is our enemy? No, not at all. We have to collaborate, we have to communicate but if you think about this in terms of control and influence, we have no control on this relationship, only influence. It is a very dynamic relationship and it has to be that way. That is the relationship I look for … we have to understand that the media have multiple roles and that I hope when issues arise we are in a collaborative mode, crucify me later, I am happy to deal with issues and questions later, but no amount of friendship would stop the media asking those questions nor should it.

The important issues here are the problems which can arise when agencies attempt to deliver information to the public during fire incidents. While the DSE and CFA co-operated in fighting the Wilson's Promontory fire and were seen by journalists as doing all they could to brief them on the status of the fire, DSE continued to be perceived negatively by journalists and was presented in a negative light. There are at least two explanations for the ongoing negativity toward DSE. The first was the DSE had started the fire, albeit as a controlled burn that escaped. The second could be related to ongoing perceptions of the DSE.

The media and community warnings: the role of the information officer

As well as using the media to promote the public profile of the fire services, it is perceived by both CFA and DSE personnel as an important tool for delivering specific warnings during incidents. This is expressed in the distinction made, in both DSE and CFA, between an 'information officer' and a 'PR' staffer.

> Corporate media officers write stories from a corporate perspective. We are information officers and we deliberately call ourselves information officers not media officers, because our purpose is the dissemination of information to more than just the media. So even though we look after the media, we also put information on our website and provide information to the Victorian Bushfire Information line. We send the information up the organisation to the Minster, to the Premier, across the Department and the community, and to other agencies. We are more then just media persons. That is why it is important to see ourselves as information officers where media is an important part but not the only part of what we do (DSE staff member).

The distinction between the corporate perspective and information delivery is not that clearcut. Although it was clear to emergency personnel that some information may damage the corporate impression, the immediate dangers of incorrect information was seen as the main issue. One DSE officer explained this in the following terms.

> The need for controlling the information can be an issue of public image but I think it is more about the risk to create panic or concerns. I hear on the radio that a particular town is on fire when really it is 10 kilometres away and in fact it is just someone's back fence that is burning. I mean the DSE has, everyone has, an opinion in the papers about the DSE and what we do, but this image stuff, that's not the real issue. The real problem is the potential to create the wrong story in the eyes of the community that may be impacted by the fire. Communities need good, accurate safe information. If we are mucking around with their heads and saying things which are not true they will simply not trust the information that they are getting (DSE staff member).

In this operational context the CFA and DSE have established an information unit in the planning section of the incident management team during incidents. The Australian Inter-agency Incident Management System (AIIMS) Incident Control System divides incident management into four distinct roles. One is the incident controller, who has overall responsibility for managing a particular incident, including all information about the incident. The incident's logistics, operational and planning roles all come under the direct authority of the incident controller. The management function remains the same whether the incident develops in size or complexity. This system may prove highly beneficial in generating a better communication flow between different agencies, across states and with the public and the media.

The information unit's objective is to deliver accurate and timely messages to communities threatened by a particular fire and to encourage residents to respond appropriately to the specific threats they are facing (Carson 2004). This approach challenges the traditional media liaison practices by claiming that the communication objective during an incident should not be to meet media needs: 'Communication with the media is not an objective, but a tool for achieving the objective of reaching those at risk' (Carson 2004). In this context, the COAG inquiry discussed the use of a standard emergency warning signal during incidents and proposed that bushfire threat warnings be conveyed consistently in all states and territories (Ellis et al. 2004: 152).

Observations of the CFA's information unit during incidents indicate that PR personnel are well-integrated into the unit. The fact that they have a media background enables them to understand the priorities of journalists and deliver information to the media more effectively. Although they work with prewritten safety messages and warnings, they can deliver these in ways that are responsive to media needs and sensitivities. A similar practice was adopted by the DSE's information unit. One staff member explained how this is done.

> We have full-time media support when we need it. We also monitor the media so that if something is happening we are on to that. I believe that we now understand better the importance of the media in the department. We cannot just say that we are putting the fire out – I placed this banner up in the Emergency Coordination Centre: 'Putting out information is as important as putting out the fire!' I would like people to really take that on board. We are starting to push that forward and this is something that has not ever been a top priority before (DSE staff member).

Radio, more than any other medium, is considered as the best means for delivering timely safety information.

> During fires we view the media, particularly local ABC radio and local radio stations, as the most effective and the fastest way for us to get messages out to the community. Whether these joint messages with the CFA are about people needing to start to defend their homes or what clothing they should be wearing if embers start landing, definitely radio is the most effective way and we see the media in this context as an ally. We couldn't do our job without them as far as the community safety aspect goes. The media may be viewed more negatively by the Department when the immediate threat has passed. The fire has passed and we start getting criticism on our control or suppression techniques or prescribed burning programs, or about incidents that may have occurred during the fires. It is then that we find ourselves on the defensive (DSE staff member).

Radio, particularly commercial talkback radio, can also be the source of the most damaging media coverage for the agencies, in relation to both their profile and their safety messages and community warnings. The immediacy and accessibility of radio, that makes it so effective in delivering safety messages, is also what makes it difficult to control and manage from a PR perspective.

> In a bushfire situation radio is the most important medium because it reports where the fire is, what it is doing and community warnings. Television helps in that way as well if fire is coming through in the afternoon. Radio news is good, especially if it is used to broadcast live and instant warnings but when radio starts getting into the realm of talkback or when some presenters start coming up with their own opinions that's when radio is not as effective and that's when media can be a lot more destructive than the newsroom (DSE staff member).

Current state of affairs

Emergency organisations in Victoria have responded to the challenges and opportunities offered by the mass media in several ways. First, they have redefined the media as a partner. Strategies designed to promote this approach include providing training for journalists in media organisations so that they can better understand fire behaviour and safety-related issues on the fire ground. Journalists who take part in this training are issued with media passes. Emergency organisations also provide media liaison officers who escort journalists on the fire ground. There has also been an attempt to develop media skills and understandings in a range of other DSE and CFA staff. In addition, all emergency organisations in Victoria have signed a memorandum of understanding with ABC Local Radio Victoria, making that radio network Victoria's official emergency services broadcaster. Most Victorian emergency organisations have developed strategies for managing and monitoring media messages. Where appropriate, they have centralised the control of media messages and regularly monitor and assess their media performance through reviewing the media coverage of their activities. For this purpose, emergency organisations have hired spokespersons and media staff with a journalism background and created a range of proactive media activities and a comprehensive media plan.

Conclusions and implications

What are the implications of this analysis for fire and land management agencies? Before discussing this issue, it is worth making two points. The first is that our research is based on Victorian views of the issues – in other jurisdictions the picture may be quite different. Second, some of these suggestions flow directly from our research and analysis while others have emerged after reflecting on the issues.

Given the significance of media coverage for emergency organisations, their staff, volunteers and the public, all emergency organisations might benefit from devoting resources to actively improve their relationships with the media. It is now common to monitor the content of media coverage, but there could also be some value in organisations undertaking regular research into the way different media organisations and journalists perceive their interactions with the organisations, and the actual stories and events they write about.

Monitoring the relationship with the media could be based on a specific event such as a major fire where journalists who had worked on that fire could be debriefed. Alternatively, journalists could be surveyed at the end of the fire season or at the conclusion of a major fire event such as the Wilson's Promontory fire. The results could be used to modify procedures and enhance the relationship between the media and the emergency organisations. Such activities may be carried out by specifically trained information officers whose role would include building effective working relationships with and understanding of different media organisations and individual journalists who write stories about the fires.

If these activities were conducted jointly by fire and land management agencies in Victoria, a number of benefits could accrue. First, journalists would be encouraged to view the agencies as co-operative rather than competitive. Second, the agencies might discover new ways of co-operating and sharing resources in the area of public communications.

On a more concrete level, there might be some value in designating one agency as the source of all information for the media during the immediate fire incident. This has several advantages: it might reduce duplication of effort and speed up the process of supplying timely information, and it might reduce the potential for conflicting perceptions of the situation among journalists. All requests for media comment would be channelled to the agency designated for this incident.

Finally, as noted earlier, land management agencies have complex relationships with the community. If a land management agency such as DSE created a task-specific and branded division which dealt with bushfires or bushfire communications, e.g. Forest Fire Victoria, it might create an institutional presence distinct from the parent organisation. If handled properly, the newly created division could interact with the media and the public without the complexity associated with communications directly from its parent agency.

Endnotes

1   The authors wish to thank staff from the CFA and the DSE and a range of media organisations for their participation in this study. Special thanks go to Gary Morgan of the Bushfire CRC for comments on a draft of this chapter.

2   The full report is available on the DSE website: http://www.dse.vic.gov.au/DSE/dsencor.nsf/LinkView/4DA3D897FFF20945CA2570AB0004391F367C6DC07DF8F640CA2570AB000DEC20.

3   The Esplin report about the Wilson's Promontory fire recommended that DSE should co-ordinate its public information and communication with the media better in relation to prescribed burns but it did not raise the possibility nor find any evidence of a cover-up in relation to the Wilson's Promontory fire.

Chapter 11

Preparing for bushfires
The public education challenges facing fire agencies

Douglas Paton and Lyndsey Wright

Chapter summary

Recent reports (e.g. COAG) identified that the low levels of community bushfire preparedness was an issue that required urgent action. This chapter opens with a discussion of how low preparedness can result from two separate processes. In one, low preparedness reflects the fact that although some people appear disposed to prepare, they need to be guided in doing so. In the second case, low preparedness results from people deciding not to prepare. Next, the chapter discusses how each outcome can be understood in terms of how people's interpretation of their circumstances and the information available to them results in their decision about whether to prepare. We discuss the role of negative attitudes, anxiety, inadequate knowledge and lifestyle factors as predictors of people deciding not to prepare. Next, we review how perception of bushfire risk, the relative importance of bushfires, distinguishing between uncontrollable bushfire causes and controllable consequences, coping style and sense of social responsibility predict people's decision to prepare. The identification of these predictors is used to identify guidelines for the development of bushfire public education programs.

Introduction

In communities susceptible to bushfires, household preparedness is a significant component of a bushfire risk management strategy. Household preparation for bushfires includes, for example, creating a defensible space around the home, actively managing vegetation, cleaning leaves from guttering, placing metal flyscreens on windows, screening eaves, ensuring access to water and equipment resources (buckets, mops, pumps, hoses, ladders) for extinguishing spot fires and determining residents' 'stay or go' decision. Despite Australia's history of devastating fires, the goal of ensuring sustained levels of bushfire preparedness has proved elusive and remains a significant public policy issue (McLeod 2003; Ellis et al. 2004). This makes it important to determine why levels of preparation remain low and to identify ways in which sustained levels of preparedness can be encouraged.

Bushfire public education programs rely heavily on providing information. However, people do not necessarily accept this information at face value. Their decision about preparation is based instead on how they interpret their risk and the available information. Analysing preparedness behaviour from this perspective has produced an interesting finding – although some people appear willing to prepare but need guidance, others simply decide not to prepare (Paton, in press; Paton et al. 2006a; Paton et al. 2005).

This means that, when planning public education, it is not enough to know that levels of preparedness are low. It is necessary to find out if levels are low because people have decided not to prepare, or because people need guidance to know what to do. Fire agencies need a means of differentiating between people who are predisposed to prepare and those who are not.

This can be done through assessing people's intentions (Paton et al. 2005; Paton et al. 2006a). The relationship between intentions and preparing is illustrated in Figure 11.1. One intention, 'intention to seek information', illustrates that people can be motivated to access information but do not translate this into actual preparedness. Indeed, the significant negative relationship between intention and preparedness illustrates that the stronger the intention, the less likely people are to prepare. In contrast, people who form 'intentions to prepare' are more likely to actually prepare.

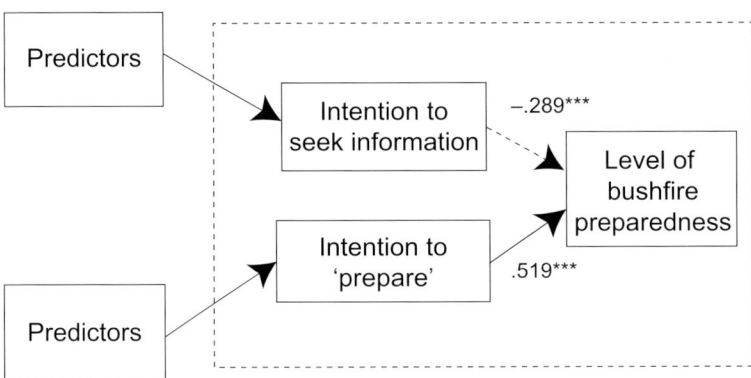

Figure 11.1   The relationship between 'intention to prepare' and 'intention to seek information', and actual level of bushfire preparedness in Canberra and Hobart, 2004–05.
Source: Adapted from Paton et al. (2006a).

Because the reasons underlying low preparedness ('decided not to' versus 'need guidance') represent different ways in which people deal with their bushfire risk, public education programs must comprise two separate strategies. The first must address the reasons that lead to people deciding not to prepare. The second provides the guidance people require to prepare. To design these strategies, it is necessary to identify the factors that predict each kind of intention. Key predictors of decisions to prepare include perception of bushfire risk, the relative importance of bushfires, distinguishing between uncontrollable bushfire causes and controllable consequences, coping style and sense of social responsibility. These issues were identified in a study of people's preparedness decisions in communities at the peri-urban interface in Canberra and Hobart during the 2004–05 bushfire season (Paton & Bürgelt 2006; Paton et al. 2006a). Full details of the locations and methods adopted are available elsewhere (Paton et al. 2006a).

Understanding reasons for not preparing

Factors contributing to people deciding not to prepare are summarised in Figure 11.2. An important predictor of deciding not to prepare is a negative attitude toward bushfire risk and preparedness. Negative attitudes were reflected in people:
- being willing to take the risk
- perceiving bushfires as having a lower priority in life than other demands
- having a low sense of attachment to property and place

- believing that insurance is sufficient
- believing that preparing will not make a difference.

Deciding not to prepare was also influenced by anxiety about bushfires, inadequate knowledge about the causes and management of bushfires, lifestyle choices and pro-environmental attitudes (Paton & Bürgelt 2006). For example, people whose lifestyle choice and environmental beliefs motivated their decision to live in or near the bush were happy to support measures with low environmental impact (sweeping up leaves, mowing the lawn). However, they were reluctant to support actions (controlled burning, felling eucalyptus trees) that would destroy the environment and lifestyle they value.

Social pressures from family and other community members that foster the belief that preparation is unnecessary or less important than other issues facing the community also contributed to people deciding not to prepare. When factors such as these dominate their thinking, people interpret their situation in ways that lead to their deciding not to prepare.

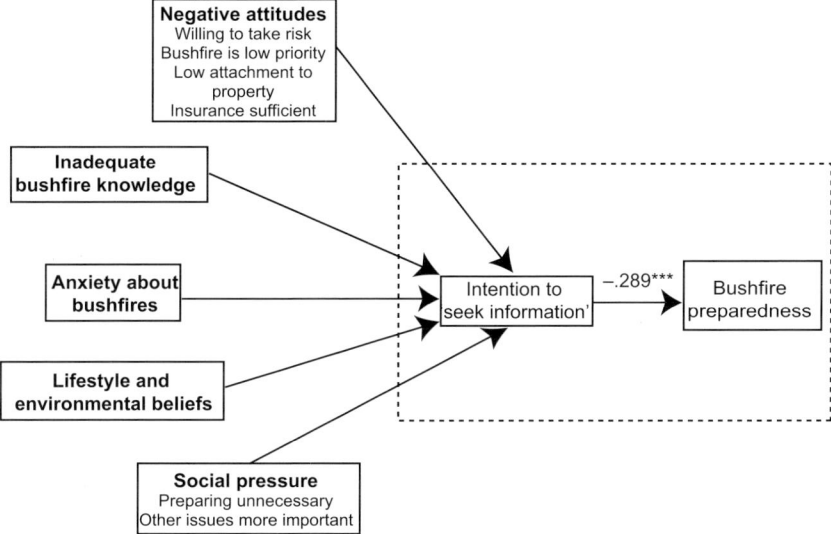

Figure 11.2 Factors influencing people deciding not to prepare.
Source: Adapted from Paton et al. (2006a).

Implications for public education

Understanding why people decide not to prepare lets us identify the issues that public education programs must address. For example, encouraging people to consider their potential personal and psychological losses can help them appreciate that they need to do more than be insured. For people with strong environmental values, demonstrating how environmental beliefs and bushfire mitigation strategies can complement one another (the relationship between burning and regeneration) can increase their support for preparing (Paton et al. 2006b).

Anxiety about bushfires is a more difficult issue to manage. If it is serious, such as when people have experienced significant loss from bushfire, clinical intervention may be required. Moderate anxiety can be managed by identifying vulnerable groups in the community and inviting people to consider what can be done to assist them. Diverting attention from people's own concerns means that anxiety is less likely to be triggered by thinking about bushfires. If people accept that their actions can help others, they are more likely to appreciate the benefits of preparing.

Preparing for bushfires

Research identified deciding to prepare as a two-stage process (Fig. 11.3). People must first be motivated to act. Once motivated, preparing is a function of whether people expect that adopting the recommended actions will lead to greater safety (outcome expectancy judgments). Preparing is also influenced by coping style (action coping) and people's sense of social responsibility.

Factors that motivate people to act

Two factors – risk perception and the perceived importance of bushfires – were identified as motivators. Risk perception reflects the degree to which people believe that a bushfire could threaten them. However, because bushfires are not the only issue facing communities, motivation is also influenced by the relative importance of bushfires compared with crime and access to health care, for example. The more important an issue becomes, the more likely it is to have a critical influence on decision-making and motivating action. It is important to recognise that risk perception and perceived importance are independent of each other; public education must address both.

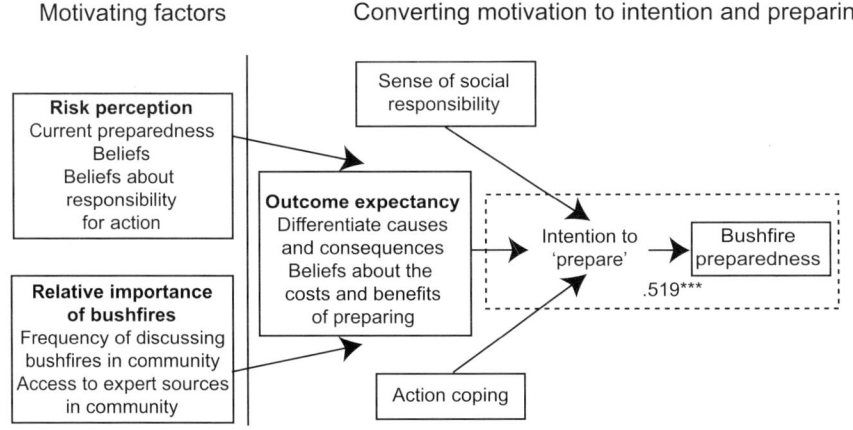

Figure 11.3   Factors that influence decisions to prepare.
Source: Adapted from Paton et al. (2006a).

Risk perception

Unless people believe that a bushfire could pose a threat to them or their property, they will not be motivated to act. Living in an area susceptible to bushfires may not be sufficient to motivate people to prepare. People's judgments are based on how they perceive or interpret risk. Several factors interact to inform people's risk perception. These include beliefs that:
  • current levels of preparedness are adequate
  • responsibility should be transferred to fire agencies and other community members.

Existing beliefs about preparedness and threat

Risk perception is influenced by people's beliefs about what constitutes adequate preparation and when protective measures should be adopted (Paton & Bürgelt 2006). Paton and Bürgelt found that beliefs regarding sufficient preparedness for bushfires ranged from just mowing the lawn regularly to implementing multiple preparedness measures. If people believe their current actions are sufficient, they are less likely to listen to risk messages or to act on recommendations.

Paton and Bürgelt also noted that although some people habitually prepared at the commencement of the fire season, others did so only when dangerous weather (hot, dry and windy) and/or bush conditions prevailed, or when fire was perceived as a direct threat (e.g. visible flames) to their property. If people's interest is triggered by proximal factors, they are less likely to attend to public education messages and may not act until it is too late for protective measures to be fully effective.

Implications for public education

Changing these beliefs involves encouraging people to personalise the threat. This can be done by inviting people to think about the personal consequences (e.g. loss of valuable belongings, family well-being, employment) they could experience. This can increase risk acceptance, but it may not change people's thinking so that they realise that they are the ones that should do something about their risk. We need to understand how risk acceptance interacts with people's beliefs about the causes of bushfires which, in turn, influences who they believe is responsible for action.

Causes of bushfires and responsibility for action

Kumagai et al. (2004) found that, irrespective of whether people had experienced a fire, most people linked the causes of bushfires to fire agencies (e.g. because fire agencies manage risk through controlled burning) (36–57%) or to natural causes (e.g. lightning strikes) (30–49%) (Table 11.1). Only 13% (on average) saw themselves as being able to influence the causes of bushfires.

If people perceive that fire agencies have more control over causes, they are more likely to transfer responsibility for managing their bushfire risk to fire agencies (Adams 1995; Kumagai et al. 2004; Paton et al. 2006a). This also means that people are less likely to act themselves. This pattern can be further influenced by people's expectation that, if threatened by bushfire, they will get the same level of response from fire agencies they would get in the case of a domestic fire.

Table 11.1 Beliefs about responsibility for bushfire causation under conditions of no bushfire experience, experience more than three years ago and bushfire experience (within the past year)

	No bushfire experience (%)	Experience >3 years ago (%)	Recent experience (%)
Actions of fire agencies	47	57	36
Nature	40	30	49
My actions	14	12	15

Adapted from Kumagai et al. (2004)

Implications for public education

To counter these beliefs, fire agencies need to engage with community members to emphasise that they work in partnership with the community and that agency mitigation strategies (e.g. controlled burning) complement – not replace – household actions such as creating a defensible space (Paton & Bishop 1996). Risk communication should emphasise that, if a bushfire occurs, resources will be stretched and there will not be enough resources to safeguard all properties. This approach can reduce the likelihood of people transferring responsibility to fire agencies.

However, a final element in the process of guiding the development of accurate risk beliefs involves ensuring that people do not transfer responsibility for preparing to other community members.

Transferring responsibility to other community members

When faced with infrequent hazards, people's risk judgments are influenced by how they compare themselves with others. For example, Paton et al. (2006a) asked people to rate how prepared they believed they were for a bushfire. Next, people were asked to rate, relative to themselves, how prepared they thought other people in their community were for a bushfire. People consistently rated themselves as being better prepared than average. This interpretive bias is known as 'unrealistic optimism' (Weinstein 1980).

The operation of this bias (everyone cannot be better prepared than the average) means that even if people accept the need for greater preparedness, they believe the information is meant for others and not themselves (because they believe they are already better prepared than others). In so doing, they transfer risk to others within their community. If all members are making similarly biased assumptions, they all believe that others need to act and their personal motivation to prepare is accordingly diminished.

Implications for public education

What can be done to counter this interpretive bias? Although it is reduced by experience of bushfire-related losses, fire agencies cannot deliberately expose communities to such losses. The bias can be reduced by providing information about what other bushfire-affected communities (with whom the target audience can identify) had done to reduce risk (Weinstein 1980). Public education should thus include information identified as coming from other communities.

The belief that you are better prepared than others tends to go unchallenged when people are passive recipients of information about risk, as is the case with mass media public education. Actively discussing risk issues increases the likelihood that people accept their risk and will act to manage it (Paton et al. 2006c). Encouraging discussion of risk among community members thus has an important role in the risk management process. It may also influence the other motivator, the perceived importance of bushfires.

Perceived importance of bushfires

Bushfire is only one risk facing contemporary communities. People also face challenges of unemployment, crime, health care and so on. Action to deal with any challenge facing a community only occurs when the issue reaches a critical level, that is, when enough people believe it is important enough to warrant their acting to deal with it.

The relative importance of bushfires influences whether people prepare (Paton et al. 2006a; Vogt et al. 2005) and the level of their support for fire agency strategies (e.g. reducing fuel loads) (Bright & Manfredo 1995, 1997). The question is how to assess the relative importance of an issue.

Relative importance is reflected in the degree to which bushfires are a regular topic of community discussion (McGee & Russell 2003; Paton & Bürgelt 2006; Vogt et al. 2005). Paton and Bürgelt (2006) found that sharing stories about bushfires among community members enhanced knowledge about local bushfire history and the reality of bushfires in the area, increased acceptance of the importance of preparing for bushfires, and provided information on how to prepare.

Implications for public education

One way in which agencies can encourage discussion is to invite community group (clubs, social action groups, neighbourhood watch, workplaces, schools and parent–teacher groups, Rotary, religious and ethnic groups) representatives to review bushfire scenarios, identify the potential challenges and determine how they might deal with them (Paton 2005). This ensures that the risk information available to communities is consistent with the values and expectations of that community (Lion et al. 2002; Paton, in press) and more likely to encourage preparing. The recently released Tasmania Fire Service DVD ('Bushfire: Prepare to Survive') could complement this process by providing a catalyst for retelling stories in ways that encourage individual and community decisions to prepare for bushfires.

McGee and Russell (2003) found that family and friends, particularly those with prior bushfire experience, were important sources of information about bushfires and preparedness activities. They also found that community members who were members of the local volunteer fire brigade were a highly valued source of fire and preparedness information. The credibility of these community leaders derived from their knowledge of the local situation and the hazard, and their ability to use that knowledge to help others develop suitable bushfire emergency plans (McGee & Russell 2003; Paton & Bürgelt 2006).

Converting motivation to intentions

Once motivated, people must convert their motivation into action. This involves people:
- identifying what they could do
- deciding whether the recommended actions will work
- determining whether they can perform these measures (Fig. 11.3).

These decisions reflect the operation of a process known as outcome expectancy.

Outcome expectancy

Risk communication involves informing people about actions whose implementation leads to the outcome of greater public and personal safety. People may not, however, take this information at face value. They interpret the information in light of their beliefs about whether they expect that a particular action will lead to the outcome of greater safety.

People's outcome expectancy is a significant predictor of preparing for bushfires (Paton et al. 2006a; Vogt et al. 2005). Low outcome expectancy can occur for several reasons. Prominent among these is the fact that people often assume that because the causes of bushfires cannot be controlled (e.g. natural causes), their devastating effects are also uncontrollable (Kumagai et al. 2004; Paton et al. 2006a). For example, natural causes such as lightning strikes, over which people perceive themselves as having low control, are often invoked to explain bushfire causation (Table 11.1).

Implications for public education

Although the causes of a bushfire might be uncontrollable, its consequences can be influenced by personal actions. Thus, public education must focus on differentiating uncontrollable events (i.e. the bushfire) from controllable consequences (e.g. reducing combustible material in the immediate vicinity of the home can limit the fire's impact), and emphasise the importance of the latter (Kumagai et al. 2004; Paton et al. 2006a).

Outcome expectancy can be increased by providing scenarios that allow people to see how they can exercise some control over consequences. For example, Turner et al. (1986) asked people if they thought anything could be done to help more vulnerable groups, such as people living in unsafe buildings (e.g. surrounded by vegetation) and the elderly. By carefully selecting the reference group that people used to make judgments about enhancing safety, people became less fatalistic and were more likely to accept that personal actions could make a difference.

For this approach to be fully effective, people require some knowledge of fire behaviour (effects of humidity, slope characteristics) and fire hazards (smoke, ember attack). Then public education must spell out how specific actions (masks, blocking doorways, screening eaves) can reduce risk from each hazard (Paton et al. 2006c). Similar links can be used to illustrate how, for example, creating a defensible space around a property and building characteristics (e.g. roof design) can reduce risk (Paton & Bürgelt 2006; Kneeshaw et al. 2004; Kumagai et al. 2004). The effectiveness of this approach can be enhanced by using examples of how these measures were successfully used in other communities. However, even if the efficacy of a measure is accepted, people may not act. This is because beliefs about expected outcomes are also influenced by people's perception of the costs versus the benefits of such actions.

Costs and benefits of preparing

People may not prepare if they believe that the costs (financial, time, difficulty) of implementation exceed the expected benefits. People tend to be more aware of the costs than of the benefits. The question is how to increase perceived benefits. Key issues include the number of items presented and how costs and benefits are framed.

Implications for public education

When presented with lists of many items, people may be overwhelmed. This reduces the likelihood that they will adopt any measures. People more readily undertake actions when presented with a limited number of choices which have multiple functions (i.e. more benefits relative to costs) (Paton et al. 2005; Paton et al. 2006c; Paton et al. 2006a).

Public education might first encourage people to act on items with multiple benefits (e.g. metal implements appropriate for fire situations are equally useful for gardening, screening eaves keeps both embers and birds out) then encourage the adoption of further actions (e.g. creating a defensible space).

Another issue concerns how the framing of costs and benefits affects risk judgments (Paton et al. 2006a). An effective strategy involves devaluing the perceived advantages of risky behaviours (e.g. not preparing because they are insured) while promoting the benefits of desirable actions (e.g. preparing increases family safety for more likely emergencies such as house fires, adopting structural measures increases the property value) and so on.

Personal and community factors

Decisions about preparing are also influenced by whether people believe they can implement the actions. Paton and Bürgelt (2006) found that being action-oriented influences this outcome. This can be measured using 'action coping' (Fig. 11. 3). Levels of action coping cannot, however, be influenced by providing information. This competence can be enhanced by inviting people to identify the problems they could encounter, determining how to deal with them, and implementing the solutions (Paton 2005; Paton et al. 2005). If communities lack the necessary knowledge or skill, fire agencies can work with them to facilitate this process.

Finally, a strong sense of social responsibility and a desire to put something back into the community predicted the formation of intentions to prepare (Paton & Bürgelt 2006). The level of these predictors is a function of the degree to which communities promote members' involvement and have the mechanisms to represent their views (Paton, in press).

Figure 11.4 Public outreach sign in the main street of Yarck, Victoria, just west of Mansfield. Photo © Bushfire CRC.

Table 11.2 Summary of issues that could be included in public education programs

Guidelines for:	Issues public education could consider
People who elect not to prepare	Encourage people to consider the personal and psychological consequences Manage anxiety by increasing perceived control Facilitate involvement in community life Link bushfire risk management with strategies for dealing with other community issues Reconcile bushfire mitigation and environmental preservation
People who need guidance to know what to do	*Risk perception* Personalising consequences to increase risk acceptance Emphasise complementary nature of fire agency and household risk management Provide risk messages that describe successful outcomes in similar communities Facilitate discussion of bushfire risk among community members
	Relative Importance of bushfires Invite community groups to review bushfire scenarios to identify and deal with the challenges Bushfire risk management facilitated by community leaders
	Outcome expectancy Increase knowledge of how specific preparation measures can reduce risk from each hazard Develop levels of preparation starting with the easiest/most general and gradually encourage adoption of more complex measures Frame information in ways that devalue risky choices and emphasise benefits of desirable actions
	Personal and community factors Develop effective coping by inviting people to identify and resolve their risk management challenges Facilitate community engagement

Conclusion

Encouraging bushfire preparedness is a significant social policy objective in Australia. How people interpret their circumstances and the available information determines whether they decide to prepare. Deciding not to prepare was predicted by negative attitudes, anxiety, inadequate knowledge and lifestyle factors. Preparing was predicted by bushfire risk perception, the relative importance of bushfires, distinguishing between uncontrollable bushfire causes and controllable consequences, coping style and sense of social responsibility. These predictors provide guidelines for designing public education programs. In addition to providing information, this chapter suggests that fire agencies should engage with community groups to facilitate preparing. By providing information, holding community-led discussions of bushfire issues and encouraging active community leadership, risk management strategies are more likely to embed 'being prepared' into the fabric of community life.

POLICY AND INSTITUTIONAL ISSUES

Painting by Mark Schaller

Chapter 12

Using program theory in evaluating bushfire community safety programs

Alan Rhodes and John Gilbert

Chapter summary

There is an ever-increasing need for evidenced-based practice in all aspects of emergency management, and bushfire community safety programs are no exception. Over the past few years the value of programs that increase awareness and preparedness among communities in high bushfire risk areas has risen dramatically. There is now a wide range of programs run by agencies across Australia and the evidence suggests these are diversifying and occurring in greater numbers each year. To date, information about whether these programs are really working has been largely anecdotal and success has quite often been determined by quantity, i.e. how many community meetings have been run, rather than a deeper understanding of how a program works and whether it is achieving its intended outcomes. The net result can be evaluations that only measure simple outputs rather than exploring more complex outcomes.

This chapter describes the use of a program theory approach to evaluating complex multi-site programs, based around a workshop or interview methodology. The rationale behind the approach, the process of developing a program theory matrix and relevant case studies help to illustrate how it can be extremely worthwhile in determining what programs work, for whom and in what contexts. It provides a deeper understanding of what can be expected from a program, and more reliable information on which to base decisions about how best to allocate community safety resources for bushfire.

Introduction

Programs are interventions intended to bring about change in a social condition or issue. Where there is an unmet need among a group of people, or another social problem that a community wants addressed, organisations often initiate programs that involve activities and the allocation of resources that are intended to address the need or ameliorate the problem.

In developing such programs, the question of whether the program will actually address the need or problem often receives less attention than a range of other personal, organisational and situational factors. There is a diverse range of circumstances that can influence the development of programs. For example, competition among organisations for funding and recognition, the enthusiasm of an individual overriding the planning and development of initiatives, or time and cost pressures. Programs can also become entrenched as 'part of the way we do business' and continue without appropriate review and revision to take account of new or emerging circumstances. The rapidly changing context in which organisations operate can also make it difficult to develop effective programs in response. If these circumstances alone drive program development, there is a risk that programs may be ineffective or unsustainable. However, identifying these circumstances does not suggest they are illegitimate or should be ignored. On the contrary, they identify some of the competing demands on program developers.

In order for programs to address a need or problem, the various circumstances that can drive program development need to be reconciled with a sound understanding of what is likely to constitute an effective program. We need an approach that can guide program development and support effective and efficient program management. Such an approach would help answer common evaluative questions:

- What problem is being addressed or what need is being met by the program?
- What is the program intended to achieve?
- How does the program work?
- Where and when does the program work most effectively?
- How do we monitor the implementation and outcomes of the program?
- How do we know whether the program will work in different situations?

The focus of program planning and development has often been on observable aspects such as the activities, resources, funding and practicalities of delivery with little attention to how the program actually works to achieve the intended outcomes. Lipsey and Pollard (1989) referred to this as the 'black box' model of a program, i.e. how the program works remains hidden. Without an explicit understanding of how a program is expected to work, it is likely to be difficult to determine whether it has worked, and if it hasn't worked then why it didn't work. Improving program delivery also becomes problematic if program processes are poorly understood or articulated. Similarly, without an understanding of how the program works it may be difficult to determine whether it can be transferred successfully to another location or meet the needs of a different group. Understanding how programs work, for whom and in what circumstances is critical to effective program development and management.

Program logic/program theory approach

An increasingly popular and useful approach to understanding programs is the use of program logic models and program theory (further discussed in Ch. 13 this volume). It is worth discussing the distinction between the two terms as it is vital to understanding how they can be used in conjunction as an effective evaluation tool. In essence, program logic is a simplified picture of how the objectives of a program could be achieved through a series of outcomes. These are frequently put together in the form of a hierarchy of intended outcomes. The limitation of this is that while a program logic identifies the anticipated or desired changes, it does not necessarily shed much light on how these changes will be brought about. A program theory is an extension of this program logic that helps to link intended outcomes of a program to the program activities and the underlying assumptions about how a program works. It helps to explain how the program elements are related and how they might work together to bring about the effects of a program. As such, developing a program logic can often be a natural first step in the development of a more comprehensive evidenced-based program theory.

This approach was adopted by Funnell (1997; 2000), who devised a program logic matrix. It is also worth noting that similar logic model approaches were conceptualised by Campbell (2005) in the related emergency management field of counter-terrorism and by ECONorthwest (2006) in a review of the National Fire Protection Agency Firewise Arcview program. In that review, a logic model was used to develop a framework for the Firewise Arcview program. It was developed through interviews with program staff, and described how the program was intended to work and the actual implementation of Firewise Arcview. The expected outcomes were largely substantiated in the results from a participant survey and in-depth interviews. Funnell's matrix works as an effective summary of a program theory and includes a hierarchy of outcomes as well as several corresponding columns that pose a series of questions:

- What are the criteria of success for this outcome?
- What program factors will determine success for this outcome?
- What 'non-program' factors will influence success for this outcome?
- What, specifically, is the program doing to address the success factors?
- What performance information should be collected for this outcome?
- How will this information be collected (e.g. what kinds of data)?

A modified version of this program logic matrix, combining a hierarchy of intended outcomes with a program theory approach that highlights how the program is expected to achieve the outcomes and in what contexts (Pawson & Tilley 1997), was developed by Program C7 of the Bushfire CRC to evaluate bushfire community safety programs. The approach examines key components of the program including:

- problem specification – the actual 'need' to be addressed by the program
- outcomes – the intended program outcomes as identified through the initial program logic development
- input/activities – the resources and components that make up the program
- mechanisms – the ways in which the desired changes are brought about
- context – the environment in which the program is implemented.

Processes to develop a program logic/program theory

There are several possible approaches to developing a program theory. The first approach is to develop the theory from existing social science theories and from the results of field studies and other investigations. The second method is to build a program theory from the knowledge and experience of those involved with the program, both practitioners and recipients of the program. Patton (1997) emphasised the value of this user-focused approach, arguing that it allows those involved in the program to test what they believe happens against what actually happens. In this way practitioners and program planners develop a greater understanding of how the program works. Chen (1990) noted that generating program theories from social science theories emphasises the value of objectivity, whereas the stakeholder approach emphasises responsiveness. The different methods could also be used together, combining existing theoretical knowledge in the social sciences, fieldwork results and input from stakeholders.

This section outlines a workshop approach that involves stakeholders such as practitioners, program managers and participants developing a program theory, and hence reflects a user-focused approach. An alternative approach to obtaining user input can be through interviews with program practitioners. This approach can often be more time-consuming but may provide greater detail and clarity about a specific program. The program theory workshops and interviews have been used to develop several program theories as part of the Bushfire CRC Program C7 research agenda, working with fire services in several states in Australia. These are presented later in the chapter to illustrate the uses of program theory in evaluation.

The workshop

The workshop involves 8–15 people with knowledge and experience of the program. One option is to involve a mix of practitioners, managers and participants to debate and negotiate the final program theory. Alternatively, separate workshops could be conducted, each comprising representatives from only one group, enabling a comparison of different perspectives on the program.

The workshop is a facilitated process using a combination of individual brainstorming and small- and whole-group discussion, supplemented with opportunities for reflective writing on experiences of the program. The workshop progresses through a series of focus questions, collecting and discussing workshop participants' ideas, discussing and negotiating these contributions to reach a consensus, or if necessary identifying points of difference. Typically the workshop takes four to six hours, with breaks. The contributions in the workshop are written up on A4 sheets and arranged on the wall to construct a matrix of the program elements as described below. The headings listed in Table 12.1 are used to make a series of columns on the wall so that the workshop discussion progressively populates the columns.

Table 12.1   Key components of the program examined in the matrix

Problem specification	Identify what problem the program is intended to address, the key themes or components of the problem and the specific relationships involved
Intended program outcomes	Participants develop a hierarchy of intended outcomes for the program, starting with the ultimate outcome and working backwards to identify and logically order the outcomes for each stage of the program
Activities and resources	In small groups, participants identify relevant program activities and resources that are mobilised to achieve particular outcomes
Mechanisms	In small groups, possible mechanisms are identified by asking participants to consider how each activity or resource might affect people's thinking or actions
Program context	The context in which the activities or resources are likely to lead to the intended outcomes are explored by asking participants to consider for whom, when and where the program works. Consideration is also given to the contexts in which the program is not leading to the intending outcomes and the possible explanation for this

Finally, participants can be asked individually to identify examples of an aspect of the program which they believe to be a success, describe that example and suggest aspects of the example which they believe made it a success story. These are written on a template. These examples often provide very clear illustrations of context–mechanism–outcome configurations, in effect mini program theories that can be used to further elaborate the program theory.

Occasionally, contributions during the workshop raise issues, make suggestions or provide comment on the program but do not contribute directly to the question currently under discussion. These contributions are 'parked' on another wall then reviewed and discussed later. They often identify more contentious matters, suggest improvements or extend ideas on how the program works, and as such provide additional valuable input as well as ensuring that all contributions are recognised and recorded.

The matrix that is built up on the wall is reviewed and any final comments are collected. This is transcribed along with the problem specification and other data from the workshop and provided to participants as a record of the workshop outcomes. The data are then available for development of a more refined program theory by integrating them with the output of other workshops, program documentation, fieldwork or social science theories, depending on the approach being adopted. The workshop provides a useful and efficient method of gathering stakeholder input in the process of developing a program theory. Participants also benefit from the experience by gaining greater insight into the program, hearing other perspectives and collectively developing a shared understanding of how the program works.

The interview approach

Interviews follow a very similar pattern to the workshops – specifying the problem, identifying the hierarchy of intended outcomes and populating the program theory matrix with information about the activities and resources, mechanisms and program context. However, the approach centres on a series of interviews with relevant key stakeholders in the program. These can be conducted on an individual or small-group basis. A rich level of detail can be obtained using this approach and it is extremely helpful to record and transcribe the discussions. Interviews also provide a greater level of anonymity, which can be useful if there are contentious issues to discuss.

Case studies

A program theory has many possible uses and the following case studies illustrate the application of the workshop approach and the different uses of the program theory developed in the process.

Community Fire Units (New South Wales Fire Brigades)

Community Fire Units is a well-established community based program conducted by the New South Wales Fire Brigades. The program is intended to provide a first line of property protection by training and equipping residents in high bushfire risk areas, predominantly in interface areas. Units receive a standard set of firefighting equipment and are trained by local fire crews in fire behaviour, use of the equipment, and safety. If the local area is threatened, the unit is activated to protect houses until the arrival of the fire brigade. Unit members then assist the fire crew and deal with mopping-up operations, allowing full-time fire crews to move to other areas.

In nearly ten years, over 300 groups have joined the program. Major fires in recent years have given the program a high public profile, with many groups being threatened by fire and successfully defending homes. The program is widely regarded as successful and has expanded significantly – more groups apply to join than can be accommodated in the program. Following the 2003 Canberra fires, the program has been implemented in the Australian Capital Territory (see Ch. 3 this volume for a detailed description of CFUs).

Although the program is popular and seen as a success there is only anecdotal evidence about community perceptions of the program and in what situations it works best. In the context of rapid growth, high community demand and strong government support to expand the program, the program managers believed there was a need to review where it fitted in the overall strategy of dealing with the interface bushfire risk. A program theory-based approach was suggested as a way to help clarify how the program worked and in what circumstances. This understanding could assist both in planning further evaluation work and in developing a strategy for the future of the program (see Ch. 3 this volume for an alternative methodology but complementary discussion).

Workshops with unit members and local station staff were conducted in three locations north, south and west of Sydney. The workshops followed the process outlined above to produce a theory of how the program is supposed to operate. This section discusses the workshop's results on two aspects of the program theory – the problem specification and outcomes, and how the program enables particular mechanisms. The implications of these findings for further evaluation and program planning are briefly considered.

The workshops identified several different dimensions of the perceived problem the CFU program is intended to address:

- the risk to life and property from bushfire
- residents' lack of awareness, understanding and capacity to deal with risk in bushfire-prone areas

- the fire service's inability to deal alone with the bushfire threat during major fires
- the lack of community involvement in dealing with bushfire risk.

The program seeks to achieve particular outcomes, and there was broad similarity in the outcomes identified in the three workshops. The program outcomes can be considered in three broad phases – group formation, acquisition of firefighting knowledge, skills and equipment, and groups maintaining the capacity to respond effectively. The program logic underpinning these intended outcomes appears to be that:

> IF residents form local community groups THEN they develop a motivation and collective capacity and

> IF group members are trained and equipped THEN they have the understanding and ability to deal with the threat and

> IF groups deal effectively with a fire threat to their homes THEN they can support fire brigade crews and release these resources for other activities

> RESULTING in reduced losses and impact and a safer community.

Having clarified the outcomes and the underpinning program logic, the workshops then identified the program activities, possible mechanisms and contexts in which the activities are more likely to lead to the outcomes. The full program theory is being developed, however, some initial considerations are briefly outlined. They suggest some of the ways in which a program theory approach can assist with evaluation and planning.

The program involves processes operating across several different levels, from individuals and households, to local neighbourhoods and stations, to the level of organisations, government and wider social pressures and trends. Achieving program outcomes depends on the activation of particular mechanisms in particular phases of the program. For example, if a group is self-initiated its formation is seen to arise through two processes: individual concern about the threat generated either through the occurrence of fires locally or major fires elsewhere, or through long-standing concern about local issues such as access. However, it is also suggested that these concerns need to find common expression among neighbours, in a context where there is interaction, a sense of belonging and community spirit. It is also critical that there be a local champion who can bring people together, and people who are willing to commit time to the program.

Mechanisms at the level of individual psychology, such as fear and anxiety about a risk, connect with mechanisms of social interaction at the neighbourhood level to enable a collective response through the formation of the group. This represents a context–mechanism–outcome configuration that provides a possible explanation of group formation. It suggests links to notions of social capital, with the implication that in communities where there are weaker associational networks, groups may be less likely to form, leaving households more dependent on fire brigades for protection (Pope 2006). In terms of program planning, this explanation suggests that different approaches may be required in different areas and measures of social capital may provide useful indicators of where groups are more likely to form.

A second example concerns the second phase of program outcomes, in which groups acquire knowledge, skills and equipment. The core of the CFU program is training sessions delivered by local fire crews. Several key mechanisms were identified such as self-confidence, trust and empowerment that enabled people to engage with the training and successfully acquire the skills. These mechanisms were considered more likely to be activated where there was a positive relationship between the fire crews and the local community, where the same personnel trained the group rather than different crews on each occasion, where the fire crew had credibility and demonstrated commitment to the program, and where the sessions provided hands-on practice and participants received positive feedback during the training. A clear context–mechanism–outcome configuration is evident in this phase of the program, where the importance of community brigade interaction and training approaches based on principles of adult learning are essential.

A number of the issues identified in the workshops related to brigade–community relations and the role of firefighters in the training, and suggest that the program theory as it is supposed to operate is not always evident. Given that knowledge and skills to use the equipment is central to the logic of the program outcomes, anything that reduces the effectiveness of this skills transfer is likely to reduce the program's effectiveness. There were significant differences in the program approach and implementation between different regions which appeared to affect the nature and extent of brigade–community interaction, and hence potentially the effectiveness of the training. Similarly, some firefighters were considered to be highly enthusiastic about the program whereas others were seen to be only fulfilling their duties or showed less expertise in delivering the training.

This brief analysis suggests the benefits of further empirical studies to investigate the extent to which the training reflects adult learning principles, the role of organisational arrangements that differ from region to region and their effect on brigade–community interaction and the delivery of training.

Street FireWise Evaluation (Blue Mountains Rural Fire Service)

Street FireWise (SFW) is a community education program run by the Blue Mountains Rural Fire Service (RFS) that involves delivering street corner meetings in communities in targeted high bushfire risk areas. Since the program was piloted in 2000 over 100 meetings have been run. The meetings are essentially scripted presentations to members of the general public with the aim of raising awareness of the bushfire risk in their area and the need to prepare for fire. They are delivered by volunteer community education facilitators from the local brigades in the Blue Mountains and supported by the Community Education Group at the RFS district office in Katoomba. After several years of running SFW, the Community Education Group was keen to gauge the effectiveness of the program and whether it was an efficient use of resources.

SFW evolved out of similar street meeting formats developed by the Country Fire Authority in Victoria and the Country Fire Service in South Australia. Over the five years that the program had been running, the content of the street meeting had been modified extensively to adapt it to the context of the Blue Mountains. The basic street meeting premise had been further modified to meet the specific needs of local brigades. The challenges in this evaluation were:
- to work out precisely what the program was intended to achieve (what the intended outcomes were)
- to identify where the program was successfully achieving its intended outcomes and where it was not
- to gain an understanding of the contexts and mechanisms conducive to the program outcomes being achieved
- to utilise this new understanding of how the program works, for whom and in what circumstances, to help with future program planning and implementation.

An alternative but related methodology to running program theory workshops was utilised in this evaluation. It involved conducting semi-structured interviews with a range of stakeholders in the SFW program. These included:
- program developers (particularly members of the Community Education Group)
- practitioners (trained FireWise facilitators)

- local RFS brigade volunteers
- members of the public who had attended SFW meetings.

The interviews were run in two stages: first, initial consultation meetings with key stakeholders to help clarify the intended outcomes of SFW (an initial program logic), and second, more extensive interviews with a wider range of stakeholders to help develop the overall program theory. Other sources of information also proved invaluable, particularly in the initial phase of developing the hierarchy of intended outcomes (see Fig 12.1). These included various documents related to the SFW program (e.g. meeting scripts, annual reports) and the results of two concept-mapping workshops run in the Blue Mountains as part of a related component of Project C7. The second-phase interviews were particularly important for capturing many of the aspects of the underlying program theory. An interview schedule was designed to elicit information about the contexts in which the program was achieving its intended outcomes and the likely mechanisms that were being triggered.

Ultimate outcomes	*Reduced impact from bushfires on communities in the Blue Mountains (fewer houses and lives lost)*
	(Formation of neighbourhood networks)
Intermediate outcomes	*Residents use awareness and understanding to develop a realistic survival plan, decide whether to stay and actively defend or leave early, and adopt appropriate preparations around their property, i.e. they become more self-reliant*
	Residents gain increased awareness and understanding of the bushfire risk and how it applies to their specific context
Initial outcomes	*SFW meetings are positively received by residents*
	Targeted residents hear about meeting, are motivated to attend and do so
	Brigades must actively participate in SFW by targeting high-risk communities and running meetings

Figure 12.1 Simple program logic for Street FireWise program represented as an outcomes hierarchy.

Overall, the findings of the evaluation highlighted the importance of context in the implementation and success of this program. People living in the Blue Mountains often refer to the 'upper', 'middle' and 'lower' Blue Mountains and this became a useful way to think about the contextual findings of the evaluation (see Fig. 12.2).

In appropriate contexts, SFW was found to have a degree of success in achieving initial and intermediate outcomes. This tended to be more likely where residents have attended two or more SFW meetings, or been exposed to other forms of bushfire community education activity. In contexts where SFW had not worked it had led either to program abandonment or program adaptation. Some of the adaptations proved successful but succession and sustainability issues have arisen. The report concluded that the attainment of ultimate outcomes could only realistically be achieved in conjunction with other community safety programs, policies and initiatives, both directly and indirectly linked to bushfires.

The Community Education Group of the Blue Mountains RFS developed a response to the report and has been able to factor some of the key findings into the future planning and development of the program. A major aspect of implementing the evaluation findings has been the careful consideration of the program logic matrix. This has provided a useful tool to consider the implications of aspects of the program logic, and the actions the group can take to enhance the SFW program. It has also helped the group to devise more effective ways of monitoring the program in future and to identify where alternative strategies are required for satisfactory community safety outcomes.

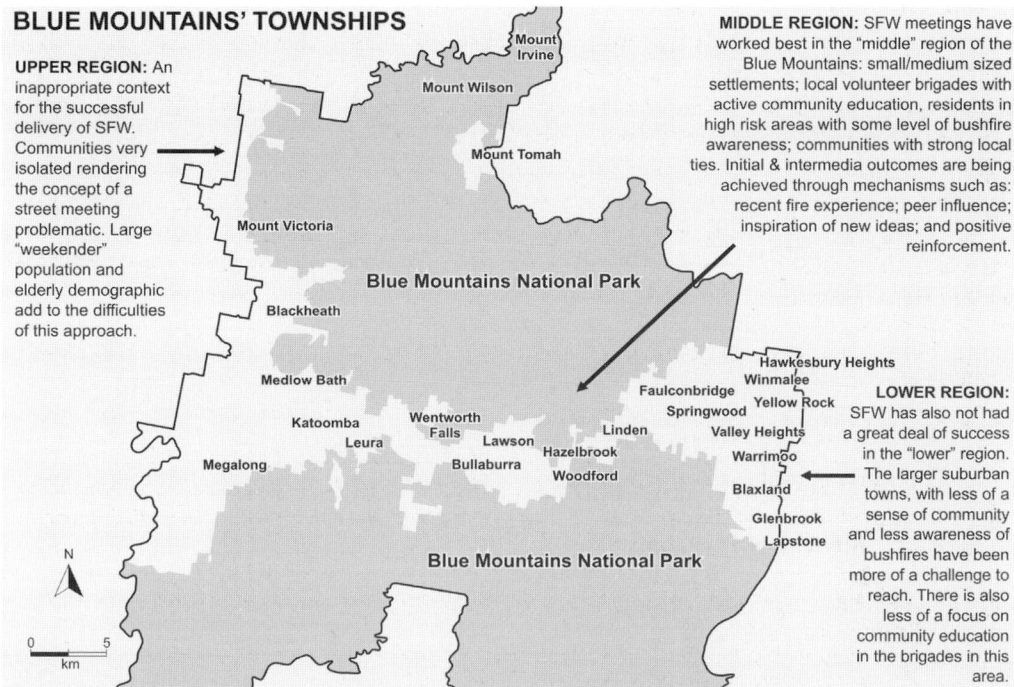

BLUE MOUNTAINS' TOWNSHIPS

UPPER REGION: An inappropriate context for the successful delivery of SFW. Communities very isolated rendering the concept of a street meeting problematic. Large "weekender" population and elderly demographic add to the difficulties of this approach.

MIDDLE REGION: SFW meetings have worked best in the "middle" region of the Blue Mountains: small/medium sized settlements; local volunteer brigades with active community education, residents in high risk areas with some level of bushfire awareness; communities with strong local ties. Initial & intermedia outcomes are being achieved through mechanisms such as: recent fire experience; peer influence; inspiration of new ideas; and positive reinforcement.

LOWER REGION: SFW has also not had a great deal of success in the "lower" region. The larger suburban towns, with less of a sense of community and less awareness of bushfires have been more of a challenge to reach. There is also less of a focus on community education in the brigades in this area.

Mount Irvine
Mount Wilson
Mount Tomah
Mount Victoria
Blue Mountains National Park
Blackheath
Medlow Bath
Hawkesbury Heights
Winmalee
Faulconbridge
Springwood
Yellow Rock
Katoomba
Wentworth Falls
Linden
Valley Heights
Leura
Lawson
Warrimoo
Megalong
Bullaburra
Hazelbrook
Woodford
Blaxland
Glenbrook
Lapstone
Blue Mountains National Park

N

0   5
km

Figure 12.2   Context is a major factor in determining program outcomes.

Conclusion

The two case studies illustrate the potential for using a program logic/program theory approach in evaluating bushfire community safety programs. Workshops or interviews involving program practitioners provide a valuable source of local knowledge about programs and their implementation. This local knowledge is particularly important in terms of understanding how programs work. Practitioners have a wealth of first-hand knowledge about how programs operate, and this experience provides insight into the way programs actually work to activate particular mechanisms in particular contexts. The two case studies illustrate the value of working collaboratively with users to develop program theories. Whether gained through interviews or workshops, this local knowledge illuminates both the mechanisms and the contexts in which programs operate and thus informs program evaluation. In addition, the program theory approach helps to highlight why a program might work in some contexts but not others. This is particularly relevant when a program is implemented in different localities and communities, as often happens with bushfire community safety programs. In these cases it is paramount to think beyond the intended outcomes to the underlying processes and contexts that facilitate their operation.

Acknowledgments

We would like to thank the New South Wales Fire Brigades and the Blue Mountains Rural Fire Service for their participation in the case studies. Thanks also to Tim Hyland, Geospatial Science at RMIT, who designed Figure 12.2.

Chapter 13

What should community safety initiatives for bushfire achieve?

Gerald Elsworth, Karl Anthony-Harvey-Beavis and Alan Rhodes

Chapter summary

Programs to increase community preparedness and self-reliance are an increasing feature of risk management for bushfire in Australia. As part of the developing community safety approach, these programs represent a significant shift in emergency management thinking. Yet at present there is little understanding of how effective the programs are, for which communities and in what particular settings they work best, or how desired outcomes are generated.

Structured concept-mapping was used in 11 workshops across five Australian states to identify clusters of ideas that describe the changes or improvements community members and fire agency personnel believe are needed to make households and neighbourhoods safer from bushfire. Synthesis of the individual workshop results yielded 14 general concepts which provided the basis for a program logic model and program theory for community safety initiatives. The logic model represents the concepts as a hierarchy of desired outcomes across a three-level view of the context within which these initiatives are developed and implemented. Concepts were also joined by linking phrases to yield a more elaborated program theory for community safety.

The results revealed how detailed and complex the notion of community safety was for our agency and community participants. Individual and household-level outcomes were seen to be underpinned and supported by a network of principles, processes and outcomes operating at the community/agency and policy/organisational levels. We argue that, to be successful, programs and policies that aim to achieve bushfire self-reliance and preparedness for individuals and households must take into account these intermediate and higher-level contextual factors that define the idea of community safety.

Insightful program design and evaluation are necessary to address a central challenge of the community safety approach – the need to develop and maintain a consistent and coherent safety message while encouraging community self-reliance, empowerment and ownership.

Introduction

In recent years major bushfires have taken a heavy toll on communities around Australia. Inquiries into these fires and recent reviews of fire prevention have addressed a common set of themes relating to improved community safety. Agencies across Australia have generally recognised that when a major bushfire occurs they do not have the resources to defend every home that may be in danger. In the past decade or so emergency management organisations have increasingly acknowledged that reducing the risk from natural hazards such as fire will be aided by the level of community preparedness and the ability of residents to respond effectively. Many emergency management organisations, including fire services, have adopted a risk management approach with a greater emphasis on prevention, mitigation and community preparedness (Smith et al. 1996). In the US, a shift in emphasis towards building partnerships with diverse communities was explicitly advocated by presenters at a national symposium of risk management practitioners and researchers in 1994 (Chess et al. 1995).

This shift in thinking from response to preparedness can be seen as a particular example of a more general policy transformation in service fields such as emergency management, criminal justice and public health towards what has been characterised as the 'community safety paradigm' (community safety approach). Defining characteristics include the general theme of protecting those at risk, securing sustainable reductions in the source of the danger and the fear of it, and an approach based on multi-agency and community partnerships (Squires 1997). Community-level engagement, responsibility and empowerment are also emphasised, and community groups are encouraged to take co-ordinated action in their own localities in association with statutory agencies and the voluntary sector.

The community safety paradigm represents a critical transformation from the approaches characterised as relying principally on the 'professionalisation of risk' and the consequent vesting of 'accountability for community safety with a professional bureaucracy' (Barnes 2002). Professionalised and expert-centred approaches include, for example, the full range of paradigms for emergency management outlined by McEntire et al. (2002). Extending from 'comprehensive emergency management' advocated by the US National Governors Association in 1979 through to the 'comprehensive vulnerability management' conception, all approaches to reducing the risk from hazards described by McEntire et al. seemingly emphasise top-down decision-making, deployment of professional expertise, and agency responsibility and control. In contrast, a central component of the community safety approach is active engagement with and empowerment of the community to investigate its own risks and develop its own solutions. In this sense the community safety approach in emergency management parallels the approach in public health that aims to realise the ideals of community empowerment and ownership of problems and possible solutions in the context of national, state and local government planning and provision of professional services (Laverack & Labonte 2000).

Reflecting this new approach to the management of bushfire risk in Australia, a safe community has been defined as 'locally organised and resourced, well-informed about local risks, proactive in prevention, risk-averse, motivated and able to manage the majority of local issues through effective planning and action' (Hodges 1999). The notion of community self-reliance is often cited. Increasingly, organisations are seeking ways to engage more effectively with communities to promote greater understanding by providing information, but also to increase community involvement through consultation and enabling communities to share in decision-making. Fire services and land management agencies now frequently advocate the importance of partnerships with other organisations, and with the community, to achieve common goals.

At present, however, there is little understanding of the effectiveness of the community safety approach to natural hazards and the specific programs designed to enhance it. If, indeed, the approach is effective it is also important to know for which households and communities and in what particular settings the programs work best, and how. This chapter describes an initial step in the development of a comprehensive framework and methodology for evaluating the broad range of community safety policy and programs from a theory and evidence-based perspective (Pawson & Tilley 1997; Pawson 2006). We describe how a workshop technique called structured concept-mapping was used to generate ideas about the wide range of outcomes that might result from the community safety approach to bushfire, and how these outcomes were arranged into a program theory about the ways in which the community safety approach might work.

Program logic models and program theories

> Program logic models and program theories increasingly form the basis for designing program evaluations in diverse service fields. Typically, a program logic model is defined as 'a picture of how your organisation does its work – the theory and assumptions underlying the program. A program logic model links outcomes (both short- and long-term) with program activities/processes and the theoretical assumptions/principles of the program' (W.K. Kellogg Foundation 2004).

The idea of a logic model grew out of an early recognition in the evaluation and public health literature that a program could be represented by an ordered series of objectives, from the idealised (long-term) objective to lower-level administrative tasks and their immediate products (Suchman 1967). From this perspective, program evaluation was seen as a process that worked up the hierarchy of objectives, determining whether each, in sequence, had been met. An evaluation of the entire program was accomplished by addressing the final long-term objective. Program logic models of this form are often represented as a diagram in which boxes in a linear sequence representing the hierarchy of objectives (or anticipated program outcomes) are linked by single-directional arrows.

More recently, it has been proposed that a program itself is a theory of social change and that, in addition to establishing that a hierarchy of desired outcomes has been achieved, an evaluation should seek understanding about the way a program operates to bring about these outcomes. Thus a program theory might be seen as an elaborated logic model with ideas about the strategies and underlying causal processes through which a program generates its outcomes, together with the contexts that best facilitate their operation. From this viewpoint, developing a theory of how the program is expected to operate is a necessary first step in any program evaluation:

> Social programmes are theories incarnate … Programmes are only as good as the theories built in to them. Programme evaluations comprise tests of programme theories. The theories that constitute programmes are, however, often unstated … Before programme evaluation is possible, therefore, it is necessary to bring these theories to the surface and to articulate them (Tilley 2004).

One general way to surface and articulate (reconstruct) program theories is to select from a variety of approaches to generating mental models or cognitive maps of program processes and outcomes working with program staff and/or recipients (Leeuw 2003). Rhodes and Gilbert (Ch. 12 this volume) describe how structured workshops and individual interviews can be used to develop mental models and program theories of existing or planned community safety programs. A small number of recent studies discuss the possible use of structured concept-mapping (see below) as another source of mental models. For example, Yampolskaya et al. (2004) gave an extensive account of the method used in generating a program logic model from a concept map in their study of a (US) state-wide community-based agency providing mental health services for children with multiple needs. They described an iterative three-step process:

- the evaluation team filled out a pro forma logic model diagram using the results of a structured concept-mapping workshop with program staff
- this model diagram was reviewed in a discussion with the staff and minor modifications were made
- the evaluation team undertook a final review and the result was checked with program staff.

A central element of the resulting logic model was a set of four categories of service and lists of associated activities that were directly based on the results of the brainstorming and clustering activities conducted in the concept-mapping workshop.

One evident strength of the mental models approach to program logic development, shown in the Yampolskavo et al. study, is that the activities and implicit theories of practitioners are a potentially rich source of ideas and hypotheses about program processes and the ways these might usefully be classified and linked. As the present study yielded multiple concept maps that were subsequently consolidated into a single list of constructs by the project team, an approach to program logic development was evolved that did not (at this stage) necessitate further work with the community and agency groups that provided the initial concept-mapping data. Additionally, an attempt was made to represent the nature of the causal processes implicit in the arrows linking the concepts in the logic model by seeking to identify the most appropriate linking word (or phrase) for selected pairs of concepts. In this way we were able to move beyond a simple hierarchy of objectives logic model towards a more elaborated program theory. Linking words are a critical feature of the freehand concept maps generated in one-on-one interviews with school students, for example, to reveal individual understandings of complex concepts (Novak & Gowin 1984) but they appear to be rarely used in program logic models in the evaluation literature. For a very interesting example, see McClintock (1990).

Structured concept-mapping

The term 'concept-mapping' can be applied to any process that results in a diagrammatic representation of the way an individual or group thinks about the content and relationships associated with a specific object, idea or issue. The method of structured concept-mapping used for the current research is based on the work of William M.K. Trochim and was assisted by Trochim's computer program, the Concept System. This methodology is particularly suited to work with groups of participants developing a conceptual framework as a guide for program planning and/or evaluation.[1]

Table 13.1 Distribution of participants across the 11 concept-mapping workshops

State	Community group	Bushfire agency
New South Wales	2 workshops: 6 or 7 participants	2 workshops: 5 or 6 participants
South Australia	1 workshop: 10 participants	1 workshop: 12 participants
Tasmania	1 workshop: 8 participants	1 workshop: 7 participants
Victoria	1 workshop: 10 participants	1 workshop: 9 participants
Western Australia	–	1 workshop: 6 participants

Fire agency personnel and community members who were part of local bushfire safety groups took part in 11 concept-mapping workshops. Table 13.1 shows the location and the nature and number of participants in each group.

At the start of the workshop, participants were asked to brainstorm ideas in response to the statement:

> Thinking as broadly as possible, generate statements that describe specific changes or improvements you think need to be achieved to make households and neighbourhoods safer from bushfires.

The brainstormed statements were printed onto individual paper slips and returned to participants, who were asked to sort them any way they chose into piles that 'made sense to you'. Participants were also asked to rate each statement on two five-point scales according to the importance of achieving the change or improvement and the perceived difficulty in implementing it.

During a break in workshop proceedings, statistical analysis of the results of the sorting was conducted. This analysis mapped the statements onto a two-dimensional array (the point map) – each statement was represented as a single point and points that were closer together were perceived by the group as closer in meaning. Clusters of statements similar in meaning were then identified statistically (the cluster map). The resulting number of clusters was typically about one-fifth the number of statements. Participants were given copies of the point map and the cluster map, as well as a list showing the statements that made up each cluster. As a group, they then named each cluster, suggested any alterations they felt would be appropriate, and were encouraged to add their interpretations to the map and to note anything that they felt was missing from the final representation.

In order to achieve a more precise representation of the workshop results than that available from the Concept System program, data were re-analysed using more specialised computer software (Clustan Graphics: Wishart 2004). The Clustan Graphics analysis resulted in a three-dimensional cluster map for each workshop. Each member of the research team independently examined all the cluster maps, naming both the dimensions and the clusters. A consensus on the best representation of the results of each workshop was reached in a series of group meetings after the individual work.

A final meeting of the research team achieved a synthesis of the concepts developed in the workshops. Cluster names were written on sheets of paper and displayed on the meeting-room walls. The group started by pairing the cluster names that were most similar in meaning, justifying each pairing as it was suggested, and referring back to the detailed content of the clusters where necessary. After a small number of pairs was established the group worked in a hierarchical fashion, adding cluster names to established pairs or forming a new pair where appropriate. When all individual cluster names had been included in a synthesised group, each was named and a final revision was undertaken.

Results

Table 13.2 shows a typical list of concepts developed by one of the community groups and Figure 13.1 shows a cluster map from the same workshop generated by the Concept System program with additions made by the group during subsequent discussion. Seven clusters of statements were identified and named by participants. The group subsequently identified two broad regions of their concept map (Education, information and advice; Preparation of your household and neighbouring households), suggested that Clusters 5, 7 and 6 were causally related to Clusters 2, 3 and 4, and noted that the workshop process had not suggested any specific implementation strategies that might be used to achieve the outcomes identified in the map.

Table 13.2   A typical list of statements generated by a workshop group

1	Effective communication at times of a fire or a high fire risk (e.g. radio)
2	Local groups that can check individual household preparation and encourage proper preparation (e.g. at a street level)
3	Local resource people who can provide advice to others on practical things they can do to be better prepared
4	Community are educated to understand the benefits of being fire-safe
5	People understand the impact that not being fire-safe can have on them and the community
6	People at local (e.g. street) level receive advice and support from fire services about how to make their properties as safe as possible

Table 13.2   continued

7	People need to be clear about whether they will 'stay or go' based on a realistic understanding of their own circumstances and capacity
8	People are provided with clear information about things they need to consider in deciding to stay or go
9	Households have appropriate fire and evacuation plans
10	Households that have decided to stay need a readily available list of things they need to do in case of a fire
11	People need to know safe places to go to in case they need to evacuate
12	People need to see how quickly things they do wrong can lead to devastating results (e.g. through graphic television ads)
13	People need to understand the circumstances under which they can be directed to leave their property and by whom
14	People have an independent water supply and means of pumping water if there is a power failure
15	People clear rubbish and leaves etc. from their property
16	Where guidelines are issued to households, they need to be appropriately specific about exactly what they need to do (e.g. pumps)
17	People should be able to get an assessment of their property and situation, and get recommendations specific to them
18	Councils provide information about importance of cleaning up when they inform residents of their collection services
19	Local brigades and planning bodies can help residents access the tradespeople and services they need in order to be properly prepared (a one-stop-shop)
20	There needs to be a register of people with special needs in case of a fire (e.g. elderly, disabled)
21	Residents know about people with special needs in their street/locality (e.g. elderly, disabled)
22	Adequate fuel reduction in all properties in a street
23	People know about the rural fire service website and are able to use it (and the website is kept active and up-to-date)
24	Communication during a crisis needs to be less haphazard and more locally precise (e.g. using local radio)
25	Need to have efficient ways for communities to provide information about fire behaviour to the fire services
26	Better two-way communication during a fire
27	In the case of a fire residents need to feel that lines of communication within the fire services are effective (and not blocked by internal squabbles)
28	Need means of contacting owners of holiday properties to notify them of needs regarding clean-up and preparation
29	People are provided with education in their own setting (e.g. street meetings)
30	Better understanding of what neighbours have for fighting fires
31	People maintain all the equipment that they need for fire protection
32	People maintain all the equipment that they need for fire protection
33	Agencies provide positive education (benefits of being fire-safe) rather than negative education (don't do this)
34	Information/education needs to focus on practical issues that people may not know

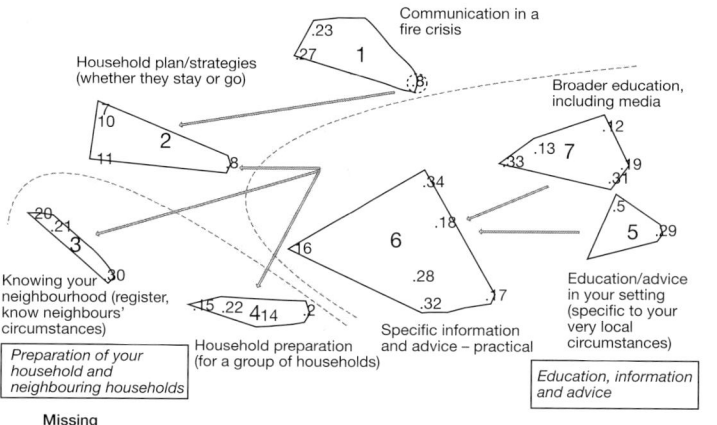

Figure 13.1   Cluster map generated during a concept-mapping workshop with a community group.

Reanalysis of the card-sort results for this particular workshop using Clustan Graphics also yielded seven clusters but suggested a slightly different grouping, with a new cluster involving residents with special needs. As well, two clusters relating to educational activities (5 and 7, Fig. 13.1) essentially merged into one named 'Community education to improve individuals' understanding and knowledge'. This process of reanalysis and comparison with the original cluster solution was followed for all workshops.

Synthesis of the results from the 11 workshops yielded 14 general concepts. Twelve were derived from the results of both community and agency workshops, two were derived from community workshops only and one from agency workshops only. Thirteen of the 14 general concepts were derived from the results of more than one workshop. The 14 concepts together with a summary of their content are listed in Table 13.3.

The ratings of 'importance' and 'difficulty' that each participant assigned to each statement

Table 13.3   A general logic model for bushfire safety programs

Concept number	Concept name	Concept description
1	Agency/community interaction	The flow of information between agencies and the public, before an incident occurs, with the aim of increasing resident awareness of the risks posed by bushfire as well as encouraging preparation to mitigate those risks
2	Household/ neighbourhood planning and preparation	The formulation of a plan that outlines an appropriate response to a bushfire and preparation that enables the chosen plan to be implemented
3	Deciding and planning for 'stay or go'	Understanding the issues surrounding the 'stay or go' message as well as making decisions about what individuals or households will do when threatened by bushfire, based on accurate information
4	Use of incentives to achieve preparedness	The use of incentives to encourage preparedness or, conversely, the use of penalties to discourage inappropriate or risky behaviour
5	Understanding/ application of regulations for bushfire safety	The need for appropriate legislation to be in place and enforced as well as ensuring community members and local governments understand why those laws are necessary

Table 13.2   continued

6	Policy framework for agency and organisational roles	Ensuring that fire agencies implement appropriate policies and procedures to support community safety initiatives
7	Principles underpinning program development and adult learning	The importance of creating an environment conducive to effective learning by adults
8	Individuals/community have a realistic understanding of risk	The focus of the statements in this cluster is on the importance of community members understanding the range of factors that influence risk
9	Appropriate information/ education activities	The provision of education to a range of groups using a number of different methods
10	Greater community ownership and responsibility for bushfire safety	The statements in this cluster are about community members taking increased responsibility for their own safety, planning for themselves and the communities they belong to
11	Agency/inter-agency responsibilities and co-ordination	Researchers identified two related yet distinct concepts. The first relates to agency responsibilities for the community. The second relates to the intra-agency relationship between the operational branches of an agency and those concerned with community safety initiatives
12	Effective communication of information during bushfire	Most statements are concerned with the way in which fire agencies deliver information to community members during a bushfire. Another element expressed in cluster 12 is that, to improve community safety, there must be systems that enable community members to communicate information to fire agencies, making use of local knowledge
13	Neighbourhood and community networks and partnerships	Most people belong in some way to community networks. These networks influence the capacity of communities to self-organise and to work effectively with fire agencies and other authorities. The networks also influence community resilience and sustainability of community safety efforts
14	Community and agency responsibilities to address specific needs	Statements are related to very specific local issues, offering practical solutions to identified problems

their workshop generated were averaged within each relevant concept. Scatterplots of these average ratings for the community and agency workshops separately are shown in Figure 13.2. There are notable similarities but also some marked differences in the ratings of the two groups of workshop participants. Concepts rated as having above-average importance in either group are located in the two upper quadrants in the plots in Figure 13.2. Concepts in the upper left quadrant are those accorded lower-than-average difficulty ratings. These concepts might be thought of as representing areas of action where more immediate achievements might be possible, while those located in the upper right quadrant might be regarded as requiring more detailed and longer-term planning.

Both groups appeared to view the concept named 'greater community ownership and responsibility for bushfire safety' as the most necessary change. The workshops with agency personnel also viewed elements of the concept 'individuals/community have a realistic understanding of risk' as highly important, but rather more difficult to achieve. In contrast, the community workshops accorded understanding of risk relatively low importance. The agency workshops also rated elements of the concepts 'household/neighbourhood planning and preparation' and 'neighbourhood and community networks and partnerships' as relatively important. 'Household/neighbourhood planning and preparation' was rated quite difficult to achieve. In comparison, the community workshops rated elements of the concepts 'policy framework for agency and organisational roles' and 'agency/community interaction' of high importance. The community workshops saw a 'policy framework for agency and organisational roles' as particularly difficult to achieve. Interestingly, agency groups did not appear to place a high level of importance on the concept 'deciding and planning for "stay or go"' while, on average, the community groups ranked this cluster of ideas fifth in importance.

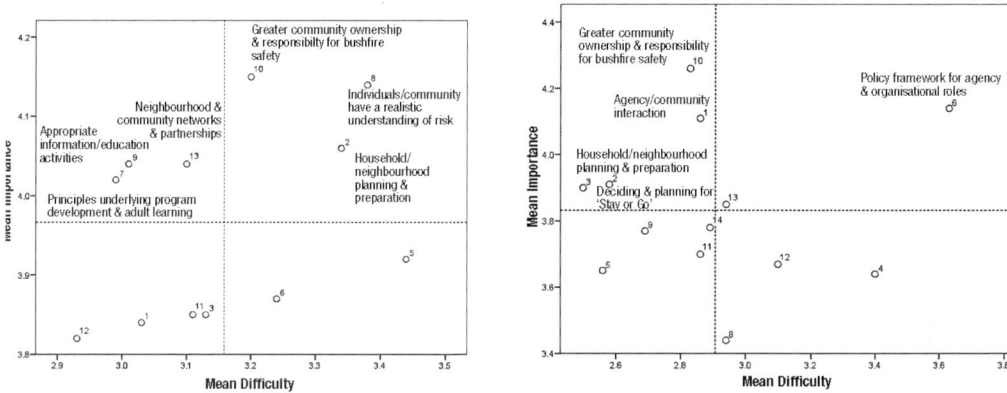

Figure 13.2   Average importance and difficulty ratings of the 14 community safety concepts.

Other concepts accorded above-average importance by the agency workshops were 'appropriate information/education activities' and 'principles underpinning program development and adult learning'. This appears to indicate a specific agency concern with the tasks involved in the planning and delivery of community safety programs.

An important feature of the 14 generic community safety concept clusters that became evident during their synthesis from the results of the individual workshops was that the concepts extended across at least three levels of desired change: individual and household; locality, community and local bushfire agency; and central agency and policy institution. These three levels, together with the time-ordered elements of an extended program logic model (context, strategies and outputs, short-term to longer-term outcomes) were used to form a two-dimensional matrix and the 14 generic concepts were sorted into the matrix cells. Next, pairs of concepts that potentially represented strong immediate (and more distant) causal links were selected (the causal arrows added by some of the workshop groups to their concept map were a useful guide). Searching the content of the statements that were encompassed by the two generic concepts for which the link was hypothesised then suggested possible linking words or phrases that best represented the nature of the causal relationship. The precise meaning of the linking words was checked against the definitions and synonyms provided by the on-line lexical reference system WordNet (Fellbaum 1998). If necessary, the most appropriate synonym provided by this system was chosen to represent the link. Finally, all members of the project team reviewed the resulting logic model (Fig. 13.3).

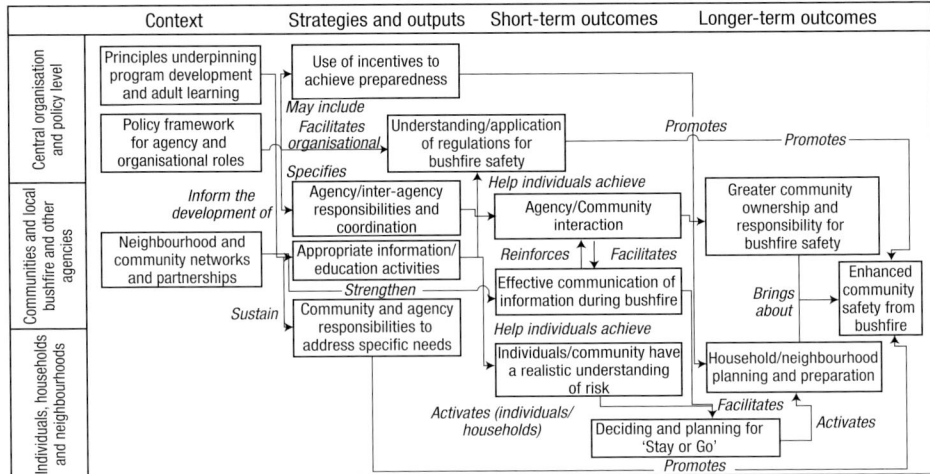

Figure 13.3   A general logic model for bushfire safety programs.

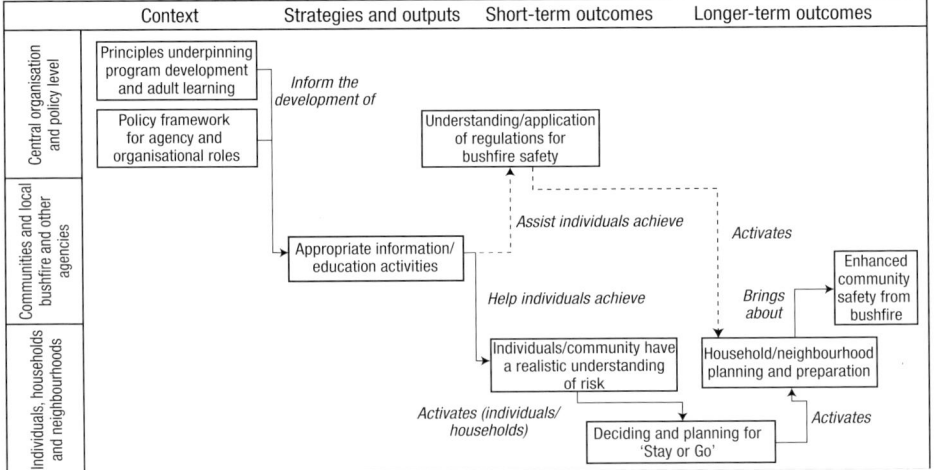

Figure 13.4   A more specific logic model for bushfire community education programs.

Implications for community safety program planning and evaluation

Fourteen generic clusters of desired outcomes resulted from the synthesis of 11 concept maps created during workshops held with bushfire agency personnel and community groups. Mapping the clusters across three organisational levels and arranging them into a provisional causal sequence with linking words and phrases provided the outline of a general program theory model that clearly reflects the community safety perspective on bushfire. In interpreting these results, it is important to appreciate that the 14 concepts are the product of a structured process that elicited then combined the ideas of 86 agency personnel and community members from five Australian states. All participants, in different ways, were closely engaged in promoting bushfire community safety.

At present, the project team is working on the application of a theory-based approach to evaluating community safety for bushfire in relation to five broad kinds of initiative:

- community education programs

- community development programs that might utilise, or seek to develop, existing community infrastructure and strengths
- media-based community safety campaigns
- regulatory initiatives, frequently requiring inter-agency partnerships
- more specific initiatives for people with special needs.

A generic program theory model such as that shown in Figure 13.3 can be used as a starting-point for more detailed theories that might underpin each of these types of initiative.

Figure 13.4, for example, shows a portion of the generic model that might usefully form a starting-point in the development of a theory of community education programs for bushfire. Using the language developed by Tilley (2004) to distinguish the intended (STD – 'supposed to do') from the alternative, possibly unintended (OAD – 'otherwise/also does') causal chains activated, the STD pathway for a community education program might be represented by the following links: 'principles underpinning program development and adult learning' and a 'policy framework for agency and organisational roles' together inform the development of 'appropriate education/information activities' which, in turn, help individuals achieve 'a realistic understanding of risk'. Realistic risk understanding subsequently activates 'deciding and planning for stay or go' which in turn activates 'household/neighbourhood planning and preparation', finally bringing about 'enhanced community safety from bushfire', the idealised long-term objective of the community safety approach. Use of the linking term 'activate' later in the list suggests that risk understanding and deciding and planning for 'stay or go' might be critical causal processes in the success of community education programs.

Possible OAD pathways can also be identified. For example, another result of the provision of 'appropriate education/information activities' might be that individual participants achieve better 'understanding/application of regulations for bushfire safety' which may, in turn, activate 'household/neighbourhood planning and preparation' leading to enhanced bushfire safety. Finally, while the (mostly) one-way arrows in Figure 13.4 suggest simple linear chains of activities and outcomes, it is important to acknowledge that a fully developed theoretical model for community safety approaches would almost certainly involve a dynamic system with many feedback loops. For example, while the concept 'neighbourhood and community networks and partnerships' is represented as a predetermining context factor in Figure 13.4, active involvement by residents in extended community education activities may also strengthen these networks and partnerships, leading to further positive community safety outcomes for that locality.

When asked to rate the 'importance' of achieving the specific outcomes identified in their workshop, participants valued changes at all levels across the spectrum of individuals and households in localities vulnerable to fire; their communities and local organisations; and central agency and government instrumentalities responsible for both bushfire response and broad policy initiatives. Both agency and community participants, on average, gave their highest ratings to elements of the concept 'greater community ownership and responsibility for bushfire safety'. Thus, both groups emphasised the importance of a central idea of the community safety approach. Agency participants generally also highlighted achievement at the individual, household and community levels related to appropriate risk perception, planning, preparation and partnerships. Community groups valued changes in relevant agencies and organisations and their interface with the community. An exception was the relatively low level of importance accorded by agency workshops to the specific concept 'deciding and planning for stay or go'.

The concept-mapping revealed the rich detail and complexity of the notion of community safety held by our agency and community participants. We believe that it is particularly important to appreciate that individual and household-level outcomes are underpinned and supported by a network of principles and processes operating at the community/agency and policy/organisational levels (see Fig. 13.3). To be successful, programs and policies that aim to achieve individual and household level self-reliance and preparedness must take these intermediate and higher-level contextual factors that define the idea of community safety into account. An ever-present and critical risk in focusing on the household level alone is that the problem becomes individualised and responsibility for action is shifted completely to householders and landowners.

The concept-mapping also revealed clear support for another central idea identified with the community safety approach – the development of partnerships across the levels of householders, communities and agencies. Further, while the specific 'stay and defend or leave early' message does not appear to have been accorded high importance by the agency groups who participated in our workshops, the more general idea of householder and neighbourhood planning and preparation was clearly supported.

These results highlight a significant challenge in implementing the community safety approach. From an agency perspective, it is clearly important that a consistent and coherent message of planning and preparation for bushfire is disseminated to householders and communities. The community safety approach, however, entails acknowledgment that communities will adapt and perhaps re-invent this message both to fit it to their own setting and to achieve ownership of it. While not denying the potential value of other approaches to community safety education, this analysis suggests that community education programs based on agency support of continuing bushfire safety community groups should represent one potentially successful model for achieving both message consistency and community ownership. Current programs of this kind include the NSW Rural Fire Service's Community FireWise groups, the Community FireSafe groups in South Australia, the Community Fireguard program in Victoria and the WA Bushfire Ready Action Groups.

More generally, the results suggest there is an urgent need for communities and fire agencies, working in partnership, to seek appropriate ways to bring about greater community engagement with and responsibility for bushfire safety in particular localities. At the same time, expert professional support and guidance will still be necessary to devise safety messages and strategies that are supported by the best available evaluation and research evidence.

Acknowledgments

The research described in this chapter was carried out by members of the Community Safety Evaluation Frameworks project team within Program C (Community Self-sufficiency for Fire Safety) of the Bushfire CRC. The authors wish to thank the other members of the Evaluation Frameworks team who contributed to the various workshops and group discussions: John Gilbert, Helen Goodman, Sandra Nolte and Sonia Whiteley. We also wish to acknowledge the essential support provided by the Collaborative Institute for Research, Consulting and Learning in Evaluation (CIRCLE) at RMIT University and the Institute Director, Associate Professor Patricia Rogers.

The concept-mapping workshops described in this chapter were facilitated and the results initially analysed by Roy Batterham, Plexus Consulting, Melbourne. The research team was greatly assisted by Roy's unique expertise. Version 1 of the Concept System program was used for the concept-mapping workshops.

Endnote

1   A detailed description of the method is given in Trochim (1989). This and many other papers describing the Trochim approach to structured concept-mapping are available on-line at http://www. socialresearchmethods.net/mapping/mapping.htm.

Chapter 14

The economics of bushfire management

Gaminda Ganewatta

Chapter summary

The terminology of economics states that while bushfires destroy economic, environmental and social assets, fire management activities require resources that have alternative uses. Fire managers face the challenge of managing bushfire threats through the use of different approaches with limited human and financial resources to minimise economic, environmental and social losses. Consequently, fire management is associated with choices and trade-offs as there are limits to the available fire management approaches and differences in the values of assets to be protected from bushfires. Making decisions that require trade-offs due to limited resources is the primary concern in economics.

This chapter explores the usefulness of economic concepts in making decisions towards better fire management outcomes. We show that economic methodologies can help in valuing economic, environmental and social assets affected by bushfires, in assessing the economic impact of bushfires and in allocating resources for different fire management programs.

Introduction

Fire and land managers face the challenge of managing bushfire risk using different management approaches with available human and financial resources to minimise economic, environmental and social losses from damage. Choosing appropriate fire management approaches and setting priorities on the protection of different asset types are important decisions made at various levels of operation. These decisions are associated with choices and trade-offs as there are limits to the available fire management approaches. For example, the use of fuel reduction burning is limited by seasonal weather and proximity to major urban centres and the availability of fire trucks, dozers and aircraft for fire suppression. There are also differences in the values of assets to be protected from bushfires (e.g. a house versus a shed). Making decisions that require trade-offs due to limited resources is the primary concern of economics. This chapter explores the usefulness of economic concepts in making decisions towards improved fire management outcomes.

Economics: science for decision-making

This section describes some of the advantages of economic analysis in fire management decision-making. First, it should be noted that the study of economics is centred on the unavoidable reality of resource scarcity among competing management alternatives. Individual actions are constrained by available time, current income or accumulated wealth, while social decisions are affected by limits to the resources that can be accessed by society. Given these constraints, individuals and society make choices by trading-off between competing alternatives. Economics focuses on how these choices are made by individuals and societies and how the results are distributed.

In the context of bushfire management, the choices faced by fire agencies include the allocation of resources for fuel reduction burning, community education or the addition of new capabilities for fire suppression within the available financial resources. The state government may have to choose the level of budget provision for fire management agencies while considering other social services such as health, education and transport. At an operational level, incident controllers may have to choose among fire trucks, dozers or aircraft in dealing with fire incidents, considering the effectiveness of each approach, the cost of operation and the value of assets at risk. At all these levels, choices are made by trading-off among alternatives, with available information and ultimate decisions benefiting from economic models and processes.

Limits to the available resources often compel decision-makers to trade-off among a number of competing alternatives, leaving some objectives unachieved. In economic terms the benefits forgone by choosing one course of action over another are usually regarded as the 'opportunity costs'. Opportunity cost plays an important role in economic analysis. Fire managers may decide to purchase a fire truck to improve suppression capabilities, knowing that this decision will leave less financial resources for community education. The inability to implement a community education program to minimise fire damage is the benefit forgone – the opportunity cost – for the addition of new suppression capabilities from fire trucks. People choose certain actions knowing the associated opportunities forgone, based on the value of the costs and benefits of the chosen action.

Values are usually expressed in monetary terms, thus providing a convenient means to describe the usefulness of goods and services exchanged through the market. However, economic methodologies can assign financial value based on human preferences for goods and services that are not traded in the marketplace. Contemporary thinking in economics assumes that individuals who prefer 'more' over 'less' maximise their own well-being or 'utility' by making rational choices among competing alternatives based on values. Thus, economic value is based on the ability of things to meet human needs or increase the well-being or utility of individuals.

Economic analysis can help to identify how resources are allocated for different uses (positive analysis) as well to show ways and means of allocating resources for better outcomes (normative analysis), thus enabling society to make best use of existing resources. Managing the bushfire risk is complex and involves many decisions at various levels. While some decisions are easy to make, others are complex as fire management issues are often associated with many economic, environmental and social values as well as consequences that may spread across generations. Economics can give decision-makers information based on sound conceptual frameworks that are equally acceptable for policy-makers, politicians and communities (Handmer et al. 2002). For example, fire management programs in a region may consider only the economic assets without considering the value of the environmental benefits that a society receives.

In summary, it is clear that economic analysis can help improve management decisions by providing information on environmental values in units comparable with economic assets. As Pyne et al. (1996) suggested, economic analysis is useful for fire management even in the presence of great uncertainty because the quantity of funds available is fixed or limited. Economic analysis can be applied to a number of fire management questions, including 'Is money better spent on ignition prevention or on fuel management?' 'Are lookouts superior to aerial reconnaissance?' 'What program mix is best?'

Bushfires as a concern in economics

From an economic point of view, bushfires can be seen as a source producing distinct impacts on the resource base of a society, affecting the well-being of individuals. Bushfires destroy existing capital stocks, natural resources and environmental services, affecting production and consumption possibilities. The impact of bushfire on livestock, buildings, machinery and infrastructure prevents the production of goods and services. Bushfires also damage agricultural output and reduce the value of timber resources. Replacement or repair of bushfire-affected capital stocks requires resources that would have been used in other productive purposes or consumption. Other important economic losses from bushfire include fatalities and injuries, health consequences from smoke and haze released from large bushfires and intangible damage to environmental resources and services such as water yield and quality and effects on leisure and recreation. Table 14.1 lists typical losses associated with bushfires and the estimated value of damages from the Canberra 2003 fires, obtained from various sources.

Table 14.1   Damages from bushfires (Canberra, 2003: total losses over $350 million)

Type of damage	Recent examples
Houses	Fully destroyed: 414 urban houses, 87 rural houses Partially damaged: 175 houses
Vehicles Cars, boats, motorbikes, trailers, commercial vehicles	300 cars destroyed, 111 cars damaged 151 commercials affected
Farmland	27 000 ha
Fencing	416 km
Nature reserves and parks	110 000 ha
Plantation forestry	11 000 ha
Loss for business	Tourism decline: school groups and nature tourism Loss of records, tools for many small business Changes to business priority of unaffected business
Environment	Reserves and endangered species breeding programs severely affected Water contamination after heavy rain washed off soil and seeds into dams and streams
Heritage/cultural	Stromlo observatory (icon of astronomy with one of the oldest telescopes, valued at c.$100 million)

The average annual damage costs from disastrous bushfires in Australia are estimated at $77 million (Bureau of Transport Economics 2001). This figure does not include the financial losses caused by a large number of small-scale bushfire events every year, or damage to environmental resources and services. From 1967 to 1999 bushfire accounted for 39% of the fatalities and 57% of the injuries associated with natural disasters in Australia (Bureau of Transport Economics 2001).

There are various fire suppression approaches operating in response to bushfire threat. These programs range from risk-reduction activities to active fire-suppression capabilities and readily available assistance for those affected by bushfires. All these programs use human and capital resources that could have been used for other useful activities for the society. For example, Country Fire Authority of Victoria (CFA) involvement with fire management programs in Victoria alone had an operational budget of $186.5 million for the financial year 2003/04. The CFA employs 1212 paid staff and draws on the services of 58 000 volunteer firefighters to meet its human resources requirement (CFA 2005). The Department of Sustainability and Environment Victoria (DSE) spent $34 million on fire management programs for the financial year 2002/03 (DSE 2003). During the 2002–03 bushfire season the state and territory governments spent more than $251 million on fire suppression activities. Fire authorities deployed over 140 aircraft at a total cost of over $110 million during the same fire season (Ellis et al. 2004).

As volunteers are heavily involved with bushfire suppression actions in Australia, large-scale bushfire events can draw huge numbers of volunteers from the workplace, affecting production and service delivery. Thus, employers bear the cost of fire suppression in terms of foregone production and service delivery that would otherwise have been produced by volunteers.

Bushfire events can also have serious indirect effects on regional economies in addition to direct economic losses. For example, the social and economic cost of the 2003 bushfires across Gippsland and north-east Victoria (in terms of loss of income and production) is estimated at $121 million in the first six months (Gangemi et al. 2003). Areas with a high dependency on tourism were affected by the disruption of tourist attractions, resulting in long-term consequences.

Taking tourism as an example, the loss of employment in one area due to diverted tourist flows can be seen as a gain if tourism income is attracted to another region – a fundamental principle of economic analysis of disaster impact is that it is the impacts on the economy in question that count. So gains and losses within New South Wales may not be of concern if the economy we are analysing is the whole state of New South Wales. The disaster loss assessment guidelines prepared by Emergency Management Australia show that transfers from one region of the economy to another, from a national economic perspective, can make big differences to the lives of the people in the region of concern. The unemployed may not be able to move freely to other areas due to associated socio-economic factors. In many cases, authorities may have to divert resources for recovery programs. For example, the Victorian government allocated $86 million to support recovery programs after severe fire damage in the Gippsland region during the 2003 fire season (Whittaker & Mercer 2004). The economic and social impacts of bushfire are important for policy-making which affects resource allocation.

Economic theory

Economics is based on the idea that a society that wishes to maximise the well-being of it members must use existing resources wisely. Thus, bushfire management effort should be designed to minimise the damage by considering the fact that damage is related to the level of preparedness to meet the bushfire threat. However, preparation to minimise the damage from bushfire requires the diversion of resources that could be used elsewhere. This position inevitably forces society to accept some amount of losses from bushfire since the cost of allocating resources for fire suppression beyond a certain level exceeds the social benefits (Hatch & Jarret 1985) – apart from the impracticality of attempting to avoid all losses. Attempts to allocate resources beyond a certain level are economically unjustifiable, as resources may be deployed for alternative uses that generate a better outcome for society. The benefits generated from alternative uses are sufficient to compensate the losses, so society as a whole is better off leaving the opportunity to compensate losses.

The economic benefits society derives from fire management efforts are reflected in the economic, environmental and social damage averted during bushfire events. It is usually expected that the higher the preparedness the lower the damage and thereby more benefits for society. The functional relationship of fire management benefits B(E) and associated costs C(E) associated with different levels of management effort are shown in Figure 14.1. The benefit increases as effort increases but at a decreasing rate. In contrast, fire management cost increases with additional management effort at an increasing rate. That is, the incremental cost of fire management effort rises as society attempts to achieve higher levels of protection from bushfire. The cost of fire management effort can exceed the benefits of programs beyond a certain effort level. This is shown at point X in the horizontal axis of Figure 14.1.

The level of effort a society places in fire management beyond point X is not compensated by the benefit it receives from damage avoided. Therefore attempts to prevent all bushfires are not economically acceptable. Effort levels of fire management that are less than point X generate more benefits for the cost of resources. In other words, the use of resources for fire management up to point X generates some extra benefits to society in terms of damage avoided. The benefit society generates in terms of damage avoided is highest at point Y, where the vertical difference is at a maximum between cost of effort C(E) and benefit of effort B(E). From an economic point of view, any level of fire management effort apart from level Y is economically suboptimal for a society that wishes to make best use of limited resources to maximise human well-being.

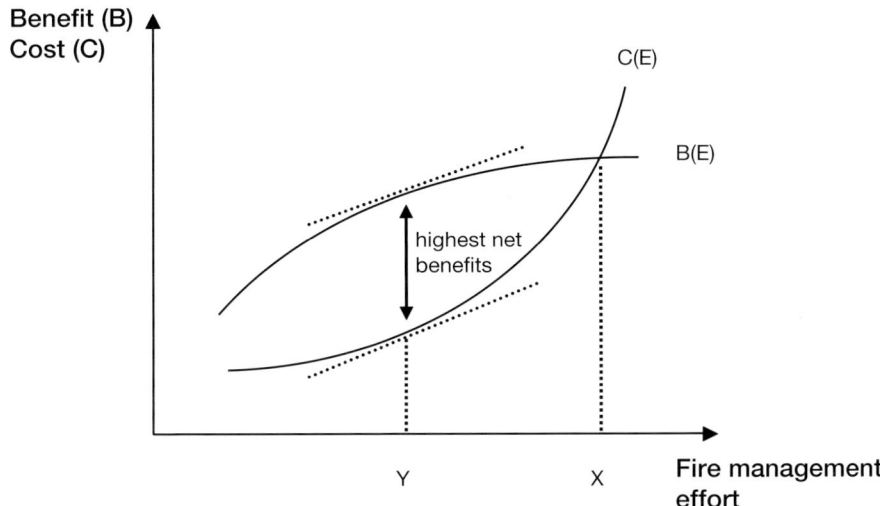

Figure 14.1   Costs and benefits of fire management against the level of effort.

The representation of fire management benefits and the associated costs of the effort provide important insights into the economic aspects of bushfire management. First, bushfire management effort beyond certain limits does not pay off in terms of the benefits received by society (note that this is subject to views about the social and political impact of being seen to do less than the maximum). Second, the choice of bushfire management effort as that which generates maximum benefits for the resources spent is acceptable for a society that has alternative uses for its limited resources.

The approach shown here provides a conceptual basis for applying cost–benefit analysis to evaluate bushfire management programs that require different effort levels. The main purpose of cost–benefit analysis is to aid effective social decision-making by facilitating the efficient allocation of resources (Boardman et al. 2001). Cost–benefit analysis enables decision-makers to prioritise alternative actions and resource allocations in accordance with the net benefits of each action. Thus cost–benefit analysis, based on sound microeconomic foundations, is a useful tool for decision-makers involved in fire management at different levels.

The damage averted by fire risk management is very difficult to identify under normal circumstances. Actual damage from fires is more visible and easier to assess. Thus, economic concepts for fire management have centred on minimising the total resource loss from the cost of fire management and fire damage. The minimisation approach is based on the idea that society desires to maximise benefits from the scarce resources available. The objective of bushfire management should therefore be to minimise the total resource cost from damage and management effort in order to maximise social benefits. The idea of minimising the total resources lost from fire provides the basis for the 'cost plus loss (C + L) and cost plus net value change (C + NVC)' criteria, frequently discussed in forest fire management. It assumes a relationship between the investment in fire protection and subsequent returns from reduced damage and control costs (Pyne et al. 1996).

Actual losses from any unexpected event are always based on a combination of the nature of the event and the state of human adjustment (Russell 1970). The economic damage from bushfire is also related to the level of preparedness to meet the bushfire threat. In particular, the level of effort society places in bushfire management decides the extent of damages at any point in time in a geographical area. Figure 14.2 presents a functional relationship of the extent of damage in terms of net value change (defined as the damage less any benefit of bushfire) from bushfire and program cost to the level of fire management effort. The extent of damage from bushfire decreases as effort increases but at a decreasing rate, while fire management cost increases with additional efforts to avoid damage at an increasing rate.

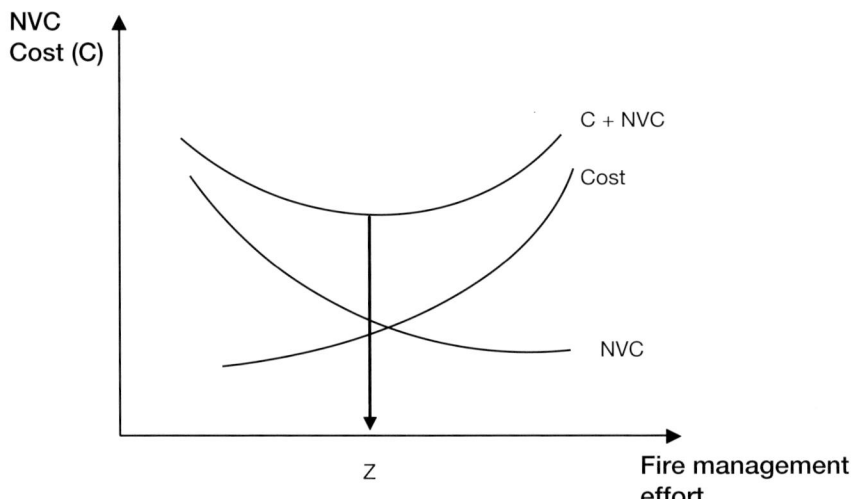

Figure 14.2   Cost of program and outcome against the level of fire management effort.

Figure 14.2 suggests that any effort to bring down damage from bushfire comes at an additional cost of resources that have alternative uses for improving social desires. So a society wishing to have the least resource losses from fire associated costs (actual damage and cost of fire management effort) should find the joint minimum of the two functions. The total resource loss from fire suppression efforts and actual damage is the vertical summation of the actual damage and cost of suppression at any level of fire management effort. Thus, society experiences minimum resource losses from bushfires at the fire management effort level shown at point Z in Figure 14.2. The graph emphasises the fact that there is an economically optimal level of fire management effort, and any effort less or greater than optimal leads to a waste of available resources in terms of damage and management cost.

The minimisation approach that provides the basis for the cost plus net value change method is logically similar to the maximisation principle which underpins the use of cost–benefit analysis (Rideout & Omi 1990). Both approaches are useful in making decisions on fire management issues for a society that wishes to maximise the well-being of its members. Thus, both analytical approaches should yield the same result in identifying economically efficient and optimal bushfire management programs.

Limitations for economic analysis

There are certain limitations to using these economic frameworks in fire management decision-making. Bushfires produce large impacts on the environment, creating intangible costs and benefits to society. Using economic analysis in the presence of a larger magnitude of intangible benefits and costs produces incomplete information. Decisions made on the basis of incomplete information may not be socially and politically acceptable. The use of efficiency arguments for the suppression decision may not allow the use of resources for the prevention of small fire events which could develop into a disastrous fire, bringing devastating economic losses if not attended early. For example, the bushfires that ran through Canberra in 2003 were initiated in remote forest areas, but use of extensive resources to contain such remote fire is not supported by economic efficiency principles. Had the initial fire event been contained, even with extensive use of resources, the massive damage caused by the bushfire could have been avoided, making the greater extent of resources allocation for initial suppression more acceptable.

Thus, resource allocation for the management of individual fire events may not be appropriate. Bushfire events could result in economic decline in regional communities due to the change of economic activities in the affected areas following bushfires. For example, many regional economies in Australia are highly dependent on farm expenditure. In the case of a large fire event, the effects on the region's farming industry would have produced serious economic impacts on surrounding businesses. In general, economic analysis does not include these effects on the regional economy. However, such changes are important concerns for political decision-making. Thus, economic analysis as a basis for decision-making may not be applicable on all occasions. Nevertheless, appropriate use of the economic framework, with its limits, will be useful in many areas of fire management.

Economics for bushfire management

The economic framework for fire management decision-making discussed above is of a general nature and can be applied across a range of fire risk management scenarios. The use of these methodologies requires information on costs and benefits of fire management actions and the value of damage from bushfires as appropriate for the intended analysis. Information on the nature of relationships or functional forms specified is also important for rigorous economic analysis based on optimisation techniques. We now explore the areas in which economic concepts can contribute to better fire risk management outcomes.

Valuing the resources affected by bushfire

Social decision-making on bushfire should be based on associated economic costs and benefits rather than the financial cost to individuals. The economic cost of bushfires includes the opportunity cost of resources used in fire management and the true economic value of the resources affected by the fire, which includes the intangible cost and benefits of bushfire.

Valuing the resources used in fire management

Resources used in bushfire management include capital and human resources largely provided by volunteers. Resources that are used for bushfire management programs could also be used to produce other social services, while the contributions of volunteers in fire management must be assigned a value if we are to represent the extent of resources used in fire management programs.

Valuing the damage to assets and outputs

Bushfires affect assets and destroy timber resources, crops, pastures and livestock products. The damage to any asset, such as buildings or machinery, results in loss of return over the lifetime of that asset. The economic value of the damaged timber stocks, crops, pastures and livestock products should be valued to represent their true economic value independent from any subsidy or taxes.

Valuing environmental costs and benefits of bushfire

The natural environment does not only provide sources of material inputs for the economic system. It also provides life support services in the form of a breathable atmosphere and a liveable climate, and a variety of amenity services including recreation and wildlife observation. They may also act as a waste receptor service (Freeman 2003). Bushfires interact with the services produced by the natural environment, affecting use and non-use values. Bushfires can also cause the release of carbon stored in forests and other forms of vegetation, causing negative environmental consequences. Conversely, bushfires produce some positive environmental and ecological effects that have economic values for society.

The environmental services affected by bushfires are intangible and values are not readily reflected through market transactions. Thus, assigning value for environmental impacts requires the use of non-market valuation approaches, most widely favoured in the field of environmental economics. The valuation of environmental damage is useful to complete a more accurate cost–benefit analysis of alternative policies, to demonstrate the relative importance of environmental consequences and to make environmental damage more obvious and comparable (Glover & Jessup 1999).

Economic impact assessment

Economic impact assessments evaluate the regional effects of actions on prices, outputs, employment and other economic factors, focusing on how those effects are distributed across a region. The assessment of a bushfire impact on regional economies provides an insight into the consequences or potential consequences of a bushfire event in an area. In this assessment, the use of economic frameworks provides a common basis for valuing impacts, providing wider acceptance by policy-makers, politicians and communities. Such information would help decision-makers design recovery programs after bushfire disasters in regional communities (Handmer et al. 2002). Information on the possible economic impacts of bushfire on regional economies can also be useful for political decision-making about new mitigation strategies for remote communities.

Allocating resources for fire management programs

The economic methodology we present can provide a conceptual framework for decision-makers to allocate resources for specific management programs, to achieve socially desired objectives. The use of these methodological tools depends on the successful estimation of economic values as shown previously.

Optimum resource allocation

The objective of bushfire management is to minimise the total resource losses from damage and fire management efforts. Enhanced preparedness for bushfire events helps to reduce the damage from destruction that comes at an extra cost for society. An economic framework can be adopted during fire management planning to identify the optimal resource allocation that helps minimise total resources loss. For example, if authorities wish to provide suppression capabilities for different regions based on the value at bushfire risk, economic concepts are useful to assign value for different assets at risk and identify appropriate allocation of suppression resources.

Providing additional resources for fire management programs

Fire management programs require additional resources for expanding their capacities to meet the challenges ahead. In these situations policy-makers face questions over the magnitude of the expected return or net saving of resources from the unit investment as a means of justifying the intended investment. An analytical tool such as cost–benefit analysis is an appropriate methodology for such concerns.

Identifying alternative approaches for bushfire suppression

Fire managers can use economic tools to make informed choices among alternative fire suppression technologies. Such decisions fall largely within the fire managers' realm, allowing greater economic concerns in their decision-making. Fire managers would also be able to optimise resource use, allowing them to provide most of the service from the available resources. Cost–benefit analysis frameworks or cost–plus– loss approaches are important analytical tools that could be used in deciding on the use of alternative suppression equipment such as fire trucks, dozers and helicopters under different field conditions. For example, our initial work on aerial fire suppression supports the use of helicopters for fires that are difficult for fire trucks or dozers to reach quickly.

Conclusion

Bushfires produce distinct impacts on the resource base of society, affecting economic behaviour. They destroy existing resources, while fire prevention and suppression programs require the allocation of resources that have alternative uses. Consequently, the presence of bushfire threats affects the economic well-being of society by altering resource allocation. This chapter has discussed how economic frameworks may improve bushfire management by the better use of the available resources.

In spite of some difficulties, the use of an economic approach is more acceptable from a social decision-making perspective as it is concerned with the net effect on society rather than on individuals or groups of individuals. The chapter has shown a number of areas where economics can contribute to fire management decision-making. Economics can provide a standard framework for valuing resources affected by bushfires, especially the intangible environmental services, to reflect their true social values. Economic methodologies can also help with better allocation of scarce resources in fire management through the assessment of major bushfire events on regional economies and resource allocation for different fire management programs.

Acknowledgments

This chapter is based on work funded by the Bushfire CRC and RMIT University. Their support is gratefully acknowledged, as is the valuable editorial support provided by John Handmer and Thomas Lowe of the Centre for Risk and Community Safety of RMIT University.

Chapter 15

Save that brigade!
Recruiting and retaining fire service volunteers to protect your community

Jim McLennan, Adrian Birch, Sean Cowlishaw and Joel Suss

Chapter summary

Australia's volunteer-based fire services all report three current headaches, to a greater or a lesser degree:

- declining brigade memberships in some (ageing) rural communities
- low levels of volunteering in some newly established urban–rural fringe communities
- difficulties in crewing appliances to respond to emergencies during business hours weekdays.

This chapter brings together the most salient findings from a program of research involving extensive surveys carried out by the Bushfire CRC Volunteerism Project team, as well as drawing upon the wider literature on volunteering.

Declining volunteer memberships are due to significant economic and demographic changes in the nature of Australian society over the last 10 years. Solutions identified and outlined to counter the effects of these trends include:

- evaluating and maintaining the well-being, efficiency and suitability of the brigade in relation to the needs of the community it protects
- investing in current members to retain their commitment and making the organisation more appealing for new members. For example, assisting with childcare, spreading workloads so that the few are not overloaded and all are included, matching assignments to the abilities and expectations of volunteers as much as possible, and organising a social program which is inclusive of members' families
- making health and safety changes within the brigade to make it more female-friendly
- publicising opportunities for self-development – career, skills, and social aspects – to attract younger volunteers
- creating guidelines and model agreements to help volunteers and their employers establish good practice arrangements for absences from work to attend fires and other emergencies.

Introduction

Globalisation of the world economy has had major impacts on the Australian economy and the nature of work:

- decline of the manufacturing and agricultural sectors
- reduction in the size of the rural workforce

- reductions in the size of government departments and instrumentalities and privatisation or corporatisation of many of the services and functions previously provided by these organizations
- increases in the proportion of the workforce which is self-employed, in part-time employment or employed under contract
- reduced income security.

The overall effect of these changes is to make it more difficult for many people to be fire service volunteers because of conflicts between the demands of volunteering and pressures to maintain an income.

Because Australia's birth rate has declined over past decades, the average age of the population is rising and the proportion of the population aged 24–45 is falling. In many rural communities, the population is ageing faster because young people move to the capital cities and regional centres to find work or to undertake study. The pool of fit and healthy individuals available to volunteer with fire services is thus shrinking. Figures provided by Victoria's CFA show that from 2001 to 2006 the median age of volunteer firefighters increased from 40 to 46 years.

In addition to an ageing population, an increased proportion of the population has come from overseas cultures. Many of these have no tradition of volunteering for emergency services. There has also been an increase in the number of single-parent families, where childcare needs make it difficult for a sole parent to be a fire service volunteer. The overall impacts of these changes make it increasingly difficult for fire services to attract and retain enough volunteers to protect communities (McLennan & Birch 2005).

This chapter is written for volunteer firefighters who occupy, or want to move into, brigade leadership and management roles: brigade captains, brigade officers, brigade secretaries and brigade treasurers. It is also a resource for staff involved in volunteer support, management and training in volunteer-based fire agencies. Members of other emergency services organisations will find the material useful for their volunteer recruiting and management activities, for example State Emergency Service and volunteer rescue, marine rescue and coastguard organisations. Finally, a range of people who work in the broad field of community development, especially work involving community protection and safety, will find the material about working with emergency services volunteers informative.

The individual studies utilised to create this chapter are summarised in Table 15.1.

Table 15.1 Research studies which form the basis of this chapter

Exit interviews and surveys of volunteers – a literature review
Recognition and service awards for volunteer firefighters – a survey of Australian agencies
NSW grain-belt community survey 2005 – a survey of households in 29 small-to-medium sized NSW rural communities
Survey of new CFA volunteers at six months
Survey of South Australian Country Fire Service women volunteers
Survey of ACT Rural Fire Service women volunteers
Age and motivation to volunteer with CFA – a re-analysis of data obtained from the survey of new CFA volunteers at six months
Survey of employers of NSW Rural Fire Service volunteers
Impacts of fire service volunteering on the families of volunteers: an agency staff perspective – interviews with 20 CFA staff involved in volunteer management and support

Note: Copies of these reports are available from Jim McLennan.

What kind of a brigade does your community need?

A volunteer-based fire brigade is not a private boys' club in a shed with shiny machines and other toys. It is made up of members of a community to protect that community against fire and related threats. Thus, the major factor which should determine the makeup and operations of a brigade is the protection needs of its host community.

The first issue is the nature of the fire threat posed to the community: grassland? forest? agricultural? rural–urban interface? structural? industrial? Threats change as land use evolves and as building and housing patterns develop. Is the brigade well-suited to meet today's threats to the community? How about the likely threats in five years?

Threats to and from the environment are only part of the picture of community fire protection needs. Who makes up the community? Are there recognisable groups of people who may be particularly vulnerable to fire danger? Potential examples include elderly residents who may be infirm and socially isolated, new residents who may be ignorant of the threats posed by fires, residents from non-English speaking backgrounds who may be unaware of fire risks and residents with disabilities. Local government agencies are a good source of information about the changing nature of communities and brigades benefit from close and ongoing links with local government and other community organisations.

How's your brigade?

The South Australian Country Fire Service developed a Brigade Health Check program which has proved very useful as a framework to help brigades judge how well they are operating. The health checklist involves several different dimensions, including:

- ability to respond and operational effectiveness
- material resources – appliances and equipment, communications capability, fire station facilities, personal protective equipment
- human resources – sufficient members to meet brigade demands now and in the future, key brigade roles filled now and in the future
- brigade leadership – now and in the future
- brigade organisation and administration
- recruitment, induction, training.
- brigade social climate – effectiveness of teamwork, member morale, conflict versus harmony, social activities for members and their families
- links between the brigade and its community – close links or disengaged? Links between the brigade and businesses and employers, community service and sporting clubs, other emergency services and related agencies such asland management, SES, local government
- brigade is broadly representative of its community – women members, a range of ages, a range of occupational backgrounds, members from backgrounds other than Anglo-Australian.

Running a brigade health check is a way of identifying the most important strengths and weaknesses of a brigade as a basis for an action plan that will build on strengths and remedy weaknesses.

Holding on to what you have: retaining volunteers

There is a well-established principle of sales and marketing – it is cheaper to retain existing customers than to attract new ones. The same is true of brigade members: it is difficult to recruit and train new members, so invest in retaining the members you have.

Who leaves and why?

Several fire agencies have surveyed former members to find out why they left. Circumstances obviously differ from community to community, but the general picture is as follows (Cyberiad 2005; Woodward & Kallman 2001).

Many who resign do so because they have no choice. Most commonly they are moving away from the area for employment or family-related reasons. There is little that can be done in such circumstances, except perhaps to write a letter of appreciation to the former member, encouraging them to join another brigade if this is possible.

Others believe that they have no choice but to resign because of age, infirmity or ill-health. In some cases these members may give up an operational role in firefighting and move into a less strenuous supporting role in the brigade, where their experience can continue to be available.

Among those who in some sense choose to leave, the most common reason is a conflict between the time demands of volunteering and work and family demands. This is particularly likely for volunteers in the 30–40 age bracket, who are likely to be married, have children of an age requiring considerable parenting, paying off a mortgage or other loan, establishing a farm, career or business, and be involved in other family-related volunteering activities. Brigades can minimise losses of these volunteers by paying careful attention to the time demands imposed on their members. Brigades can:

- roster availability for turnouts so that the willing few are not burned-out
- offer flexible training arrangements
- provide assistance with childcare
- include families in brigade social activities.

All these are potential ways to minimise conflict between volunteering and work and family life.

A significant number of volunteers who leave say they did so because their brigade was an unpleasant place to be a volunteer firefighter. They cited favouritism, not being a member of the 'in-group', feeling excluded, personality clashes, authoritarian leadership and conflict among brigade members as reasons why they resigned. Tensions between volunteers and career staff from fire and land management agencies have also been reported as a source of disillusionment with the role of volunteer firefighter.

The message is that people-management skills in the brigade leadership team are critical. In general, agencies devote considerable training resources to developing the technical and safety-related aspects of being a volunteer, but relatively little to developing people-management and leadership skills. This is an issue needing attention at regional and headquarters level. Wherever practicable, brigades should encourage and support members interested in leadership roles to take advantage of training opportunities to develop people-management skills.

Other factors which have emerged as reasons why volunteers resign include the stress of attending emergency and road accident trauma, especially when involving serious injuries or death. Most fire agencies have counselling and support services available to volunteers. Brigade leaders need to be knowledgeable about these and active in encouraging members to use the services. Volunteers' operational assignments should be matched to their abilities and expectations as much as possible.

Maintaining member commitment

Commitment is the other side of the 'why people leave' coin: it is why people stay in a brigade. The general view among human resources professionals is that there are three different aspects of commitment (Meyer et al. 1998):

- affective commitment is the 'I *want* to stay' aspect
- normative commitment is the 'I *ought* to stay' aspect
- continuance commitment is the 'I will lose something if I *don't* stay' aspect.

All three are major contributors to retaining brigade members, although their relative importance may be different for individual volunteers.

Affective commitment

Affective commitment stems from enjoying being part of the brigade. It represents the 'fun factor' in a role that often involves discomfort, inconvenience and sacrifice. Aspects of brigade life which contribute to maintaining affective commitment include:

- enjoying at least some brigade activities – firefighting, socialising with fellow volunteers, undertaking community safety and education activities, training opportunities, competitions, working with juniors or cadets, managing, administering, instructing and mentoring
- a feeling of belonging to the brigade
- being able, at least to some degree, to include work and family life in some aspects of fire brigade life, rather than feeling that brigade life always conflicts with work and/or family.

Brigade leadership can contribute to maintaining affective commitment by careful attention to:

- maintaining sufficient numbers of members so that workloads are not too heavy
- spreading workloads so that the few are not overloaded and all are included
- managing brigade workloads so that the few who live closest to the fire station do not monopolise turnout experience at the expense of other members
- encouraging members to take on special roles in the brigade for which they have an interest or an aptitude
- trying to make the brigade family- and work-friendly – flexible training arrangements where possible, assistance in arranging childcare, well-organised and efficiently conducted meetings
- equipment as up-to-date as possible
- training opportunities well-publicised.

A brigade needs a social component to contribute to brigade cohesion. The social program needs to be inclusive of the membership and members' families – it needs to be more than a program of beer and barbecues! Usually, having a social committee spreads the load and makes it easier to organise a varied social program with a wide appeal.

Normative commitment

Normative commitment is the sense of duty or obligation to continue in the role of volunteer firefighter. This sense of obligation may arise from not wanting to let the brigade down, from a concern for the community and fellow residents or from a family tradition of volunteering. Recognition is one way that normative commitment can be maintained and strengthened:

- recognition within the brigade as being a part of the team and making a particular contribution
- recognition within the fire agency
- recognition within the community as being part of an organisation that protects the community.

A brigade newsletter or similar is generally a positive ingredient of brigade life – length and frequency of circulation will depend on the size and nature of the brigade and the community.

It is usually best to assign the role of editor to an interested individual rather than load the task onto the brigade secretary. With internet access increasingly common, a newsletter or bulletin might be primarily in electronic format, with a relatively small number having to be posted to members and other parties.

The brigade leadership team should ensure that when members qualify for an agency or volunteer association award, this is communicated to the appropriate party and followed up. Obviously a good brigade records system is necessary for this.

If possible, a suitably interested brigade member should be tasked with the role of media liaison and marketing (or similar). An important component of the role is monitoring brigade member activities, arranging for photographs and feeding these plus stories to local media, both print and electronic, and to the fire agency head office.

Continuance commitment

Continuance commitment is an awareness that leaving the brigade would represent the loss of some things which are valued. This has two aspects. High levels of affective and normative commitment feed into an awareness of potential loss. Minimising the cost to the volunteer of being a bridge member is the second aspect – if the cost becomes too great, the anticipated loss will be discounted. There are three major areas of cost for volunteers:

- time devoted to volunteering is at the expense of time available for other activities, such as family and work
- opportunities to engage in desired activities may be precluded by the demands of volunteering –lost opportunities can include vacations, sporting activities or study
- financial costs – direct monetary imposts associated with being a fire service volunteer, such as petrol and telephone bills, or indirect costs in the form of forgone income because of the demands of volunteering.

Although there is a degree of inevitability about these issues for volunteers, brigade leaders can take some steps to minimise their impact. Attention should be given to brigade numbers and workload, both routine and operational, being spread across the membership rather than falling excessively on a willing few who are at risk of burning out.

A recent survey of emergency services volunteers concluded that the average volunteer is out of pocket by about $600 per annum even after reimbursements (King et al. 2006). Brigade treasurers need to make it as easy as possible for volunteers to claim reimbursements to which they are entitled.

Interviews with experienced managers of volunteers suggest that some volunteers are their own worst enemies in the sense of losing perspective about a balance between brigade life and family/work life. If families often experience negative impacts from having a member who is a volunteer firefighter they will ultimately come to resent the brigade's intrusion into family life. This may lead to the volunteer withdrawing from brigade activities, and ultimately resigning. Some actions can be taken by brigade leaders to minimise this risk:

- include family members in the brigade induction process to inform them about firefighting, including safety precautions, to prepare them for what lies ahead. Family induction seems especially necessary when the family has an urban, rather than a rural, background
- put in place a regular communication system with brigade families (especially new families) during large campaign fires. Regular updates will allay fears based on uncertainty
- have a brigade social program which includes families. Familiarity with other brigade members and their families, and brigade activities, will generate a higher level of tolerance of the demands of fire service volunteering.

Overcommitted volunteers can have detrimental effects on brigades by contributing to setting implicit brigade involvement 'standards' which most members see as unreasonable and unattainable, thus eroding brigade morale.

A resignation may have several causes

In the discussion so far, we have considered a variety of factors which may influence a volunteer to leave or continue. While it is quite possible for an individual to resign for a single reason, it should be noted that contributing factors can operate in combination. For example, a volunteer who is experiencing work- and family-related time pressures and has previously considered resigning may decide to go ahead and resign after all because of ongoing interpersonal conflicts in the brigade resulting from ineffectual brigade leadership.

Because there is a widespread awareness among agency personnel – both career and volunteer – that time pressure is a problem for many volunteers, there is a risk that volunteer resignations may be blamed simplistically on lack of time when in fact other problems concerned with (say) management of volunteers are the major causes. Figure 15.1 shows ways in which potential factors in a person's decision to continue or to leave may interact and play out.

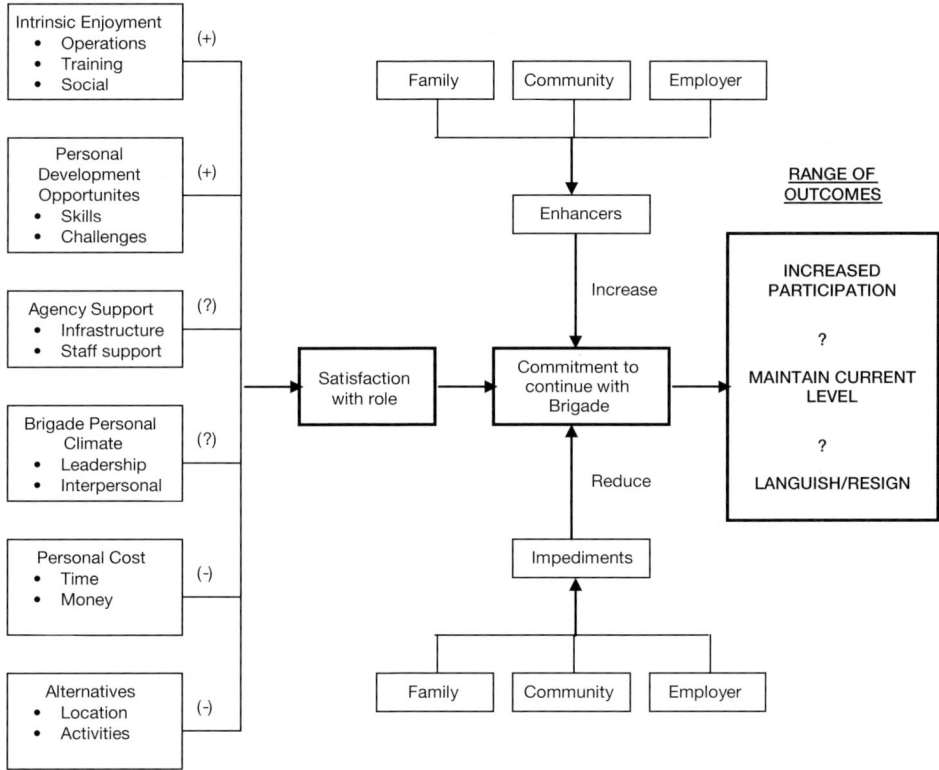

Figure 15.1 Factors which may contribute to a decision to leave or stay.

Women volunteers

Women make up 50% of the population but only 10–25% of fire service volunteers. Because emergency service agencies are traditionally dominated by males, issues faced by female volunteer firefighters have often been overlooked. This appears to have contributed to a higher than desirable resignation rate (Cyberiad 2005).

Given that a volunteer-based community protection organisation, like a brigade, will be most effective if its composition reflects the community it serves, it is clearly desirable for a brigade to include women members. The opportunity should be available for women members to train for and engage in operational duties if they wish. Anecdotal evidence suggests that brigades with female volunteer firefighters are often more open and accepting of new members and ideas, and more closely integrated with the host community, than are all-male brigades.

Brigades may have to make some changes to be female-friendly. Women are on average not as tall as men, have a shorter reach and have somewhat less grip and lift strength. These have implications for occupational health and safety issues, such as the height at which equipment is stored, the weight and usability of equipment and the operability of vehicles.

Surveys of women volunteers in the South Australian Country Fire Service and the ACT Rural Fire Service (McLennan & Birch 2006a, 2006b) found that these factors were a concern to about 35% of the women volunteers. Lack of toilet facilities at fire stations were a concern for about 25% of the women.

The level of gender-based discrimination and sexual harassment in brigades was found to be roughly the same as rates of discrimination and harassment in the general community. The role of the brigade captain in resolving discrimination and harassment complaints satisfactorily was absolutely central. Fire agencies have principles and procedures for managing discrimination and harassment complaints and issues. A brigade leadership team needs to be aware of the agency principles and procedures, and act speedily and decisively to implement them.

Recruiting new volunteers

The preceding section discussed how to ensure that a brigade is a good place in which to be a volunteer so that resignations are minimised – there is little point expending time, energy and money on recruiting new members if they are not going to stay! However, with a mobile workforce and a changing society recruiting new members will always be important.

Why do people join volunteer fire brigades?

A survey of new CFA volunteers six months after they joined showed that people become volunteers for a mixture of reasons (Birch & McLennan 2006a). The reasons cluster into three categories:

- community contribution motives
- community safety motives
- self-interested motives.

Community contribution motives are concerned with giving something back to the community – a desire to contribute, often as a kind of 'thank you' to the wider society for benefits enjoyed. Community safety motives arise from an awareness of risk to life, property and the environment posed by fires and a feeling of obligation to be an active part of the solution. Self-interested motives involve 'what's in it for me?' aspects of being a volunteer firefighter. Those identified most frequently include career enhancement opportunities, opportunities to learn new skills, meeting new friends, camaraderie and extended social contacts, and personal growth through facing new challenges.

For any individual, one of these may be more important than others. The CFA survey indicated that there are age-related differences in the relative importance of the three clusters of motivations. When volunteers were broken up into three age groups (18–34 years, 35–44 years and 45+ years), they were found to have the same levels of community contribution and community safety motives. However, the younger age group (18–34) reported higher levels of self-interested motives than did the other age groups. The implication is that three aspects should be emphasised when marketing fire service volunteering – opportunities to contribute to the community, opportunities to take an active part in protecting the community and opportunities for self-development and challenges.

Other factors and other players

Individuals' motives to volunteer are fundamental. However, other factors are important. Figure 15.2 is a simple model of the volunteering process, illustrating some of the factors and other players likely to be important in a decision to volunteer.

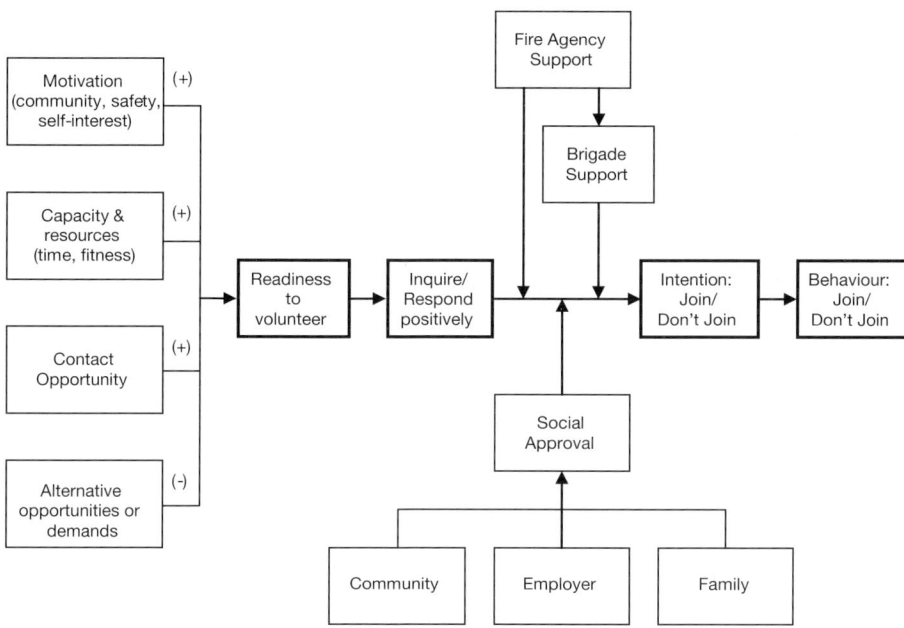

Figure 15.2 Factors in the volunteering process.

   The elements on the left are individual factors, which have already been discussed. Other relevant individual factors include personal resources (e.g. time) and capabilities (e.g. health and fitness). While these are largely fixed, they may not necessarily be barriers to volunteering. Thus, an individual may have false beliefs about the nature of the work and the risks involved, and providing factual information about what fire service volunteering actually entails may result in a more favourable attitude. Similarly, an individual may be unaware of important roles in a brigade (other than operational firefighting) where they could contribute. A recent survey of more than 1000 residents in 29 small to medium-sized rural communities in the NSW grain-belt showed a surprising level of ignorance and misinformation about the actual nature of volunteering and how volunteer fire brigades operate (Birch & McLennan 2006b). Brigade leaders should not assume that people in the local community have an accurate understanding of important aspects of being a volunteer firefighter, such as the availability of accident insurance and job protection for volunteers. They should take every opportunity to educate the community about what being a brigade member involves, via demonstrations and displays, open days and the local media.

Figure 15.3   Ungarra CFS Brigade, South Australia.
Photo © South Australia Country Fire Service.

As an extension, in order to volunteer a person must know about the need for volunteers and have contact opportunities. Information must be widely available in the local community about the need for new volunteers, and who and how to contact for more information. A noticeboard outside the fire station, entries in local government services listings, regular pieces in the local newspaper and a brigade website can all be important.

Research shows that the person likely to volunteer is likely to be interested in a range of alternative volunteering opportunities. In one sense, being a volunteer firefighter competes with other opportunities to volunteer. Leaders need to ensure that the brigade is very visible and that the work of the brigade is publicised in the local media so that fire service volunteering remains high in the public's awareness. As suggested earlier, allocating a specific brigade role of media liaison and larketing officer may be a worthwhile investment.

The elements at the bottom of Figure 15.2 represent other key players in the volunteering process apart from the potential volunteer. How positively or negatively a brigade is regarded generally by the host community is an important determinant of the chance that a community member will volunteer to join. In the survey of communities in the NSW grain-belt , there were great differences among the 29 communities about how positively they regarded their local brigade. A brigade's standing in its community is not simply a matter of publicity. It is also a matter of how strongly integrated the brigade is with community life generally. It appears that level of brigade/community integration lies on a continuum. At the positive end is a brigade whose membership reflects the general composition of the community, whose members participate actively in other aspects of community life, and which is open to and inclusive and tolerant of new members. At the negative end is a brigade which operates as a kind of private members club closed to outsiders and intolerant of differences, and whose activities are mostly unrelated to the life of the wider community.

The attitude of employers to fire service volunteering is crucial to many potential volunteers. A recent survey of NSW employers' views about employing Rural Fire Service volunteers found that most employers were accepting of employees taking time off work to attend genuine emergencies (Birch & McLennan 2006c). However, some employers did not permit employees to attend emergencies during working hours. There was no evidence that these employers were ideologically opposed to their employees being volunteer firefighters. Rather, their opposition stemmed from the nature of their business – staff absence to attend fires would have very adverse impacts on clients and customers and thus on the core business.

A small number of employers were not supportive of their employees attending fires and other emergencies even outside working hours. These were mostly in industries with stringent safety requirements, where employees needed to be alert at all times – mining, heavy manufacturing, transport. Clearly, some employment settings will be more compatible with employees being fire service volunteers than others. This needs to be taken into account by brigade leaders and by potential volunteers. Some agencies may have information packs for employers about what fire service volunteering entails, the conditions under which volunteers are afforded job protection and ways in which the agency publicly recognises employers. Some agencies are considering drawing up model agreements for employers and employees who are volunteer firefighters, setting out agreed principles and procedures for attending fires and other emergencies during working hours. Brigade leadership teams should contact their agency head office to see what information is available for volunteers to give to employers. Volunteers should be encouraged to establish an agreement with their employer about taking time off work to attend fires and other emergencies.

Families of volunteers are often key players in the volunteering process. Family members, usually a spouse or partner, can be an important source of support for the volunteer and often contribute to the work of the brigade – taking and relaying phone messages, assisting at social functions, assisting in fundraising. However, they can also be a potential barrier to volunteering if they are, for example, fearful of the risk of death or injury. Often, providing information and an informal introduction to the role of a volunteer firefighter will allay such fears.

Families can derive benefits from having a member who is a fire service volunteer. The volunteer will learn skills in dealing with fires and other emergencies which will result in enhanced protection for the family. Also, being linked with a brigade can expand and strengthen social networks in the local community, which is likely to be particularly attractive to a family which is new to the community.

The upper part of Figure 15.2 shows the fire agency and brigade components of the volunteer recruiting process. Fire agencies have a range of recruitment-related materials, and brigade secretaries need to make sure that they have adequate supplies of the most-up-to-date materials to give to potential volunteers and new members.

While fire agencies can produce marketing and information materials and arrange for television, radio and newspaper advertisements, effective recruiting of new volunteers is basically a brigade matter. Recruiting new volunteers should be purposeful, rather than accidental. A brigade leadership team needs to take into account what kinds of volunteers are needed so that the brigade's community protection activities will be as effective as possible. This involves recruiting to fill roles in the brigade, and succession planning to ensure continuity of roles. Particular attention needs to be given to recruiting younger volunteers aged 18–35. New volunteers can be recruited in two ways – they make contact with a member of the brigade, or a brigade member invites them to join. The personal approach and invitation seems to be most effective, but a brigade needs to be highly visible and easily contactable by people new to the area.

Conclusion

A brigade needs to actively welcome interested people and new volunteers, not just let them wander up and hang around on the fringes. Brigades need an up-to-date information kit for new volunteers that explains who's who in the brigade and answers frequently asked questions. Many brigades report that a mentor or buddy system for new members works well, if the mentor is the right person.

THE FUTURE WITH A WARMER CLIMATE

Painting by Mark Schaller

Chapter 16

Climate change and community bushfire resilience

Karyn Bosomworth and John Handmer

Chapter summary

Bushfire vulnerability and resilience occur in dynamic and complex ways – spatially, temporally, ecologically and socially, and in synergies between these. Along with increasing fire weather risk, climate change is also likely to affect social–ecological bushfire vulnerabilities and challenge bushfire resilience in complex ways. In this chapter we summarise projected impacts on fire weather risk, and discuss the implications for community bushfire resilience. We conclude with key questions and considerations for fire management agencies.

For communities, climate change will not only likely increase the risk of bushfire and its associated impacts and losses, but it may also affect underlying vulnerabilities and resilience. For example, combined with globalised economic pressures and drought losses, a bushfire may represent a threshold beyond which an individual or community cannot cope or recover. Less rain in certain regions may contribute to rural decline, further reducing social networks and volunteer numbers. Extended fire seasons will mean people must remain prepared and vigilant for longer. Essential services that contribute to a community's resilience may face a growing bushfire risk. Already-stressed ecosystem function and services may decline further, with concomitant impacts on social resilience. More subtle impacts may include the nuances and sophistication of preparedness advice.

Agencies will need to explore what is realistic in terms of bushfire impacts, preparedness and suppression capabilities in a changing climate, and how this will inform their planning, policies and dialogue with communities. 'Adapting to climate change' not only suggests that fire management policies may need re-evaluation, it also implicates those who interact with fire management. Recovery efforts will need to ensure that they contribute to improved resilience both to bushfires and to a changing climate more generally, rather than re-establishing vulnerabilities. Work is needed to better understand the synergies and interrelationships between climate, ecology and society; existing vulnerabilities and ways they can be addressed; whether or not our policies are effective; and whether our policies are robust enough to remain effective in our changing climate.

Fire vulnerability: the role of climate and weather

When, where, whether and how a bushfire burns is influenced by daily weather, which is a reflection of climate. The weather systems that aggregate to form regional climate determine the seasonality of rainfall in relation to the annual cycle of temperature at any location, thereby influencing fire seasonality; inter-annual fluctuations in weather and climate influence fuel loads and condition; micro- to meso-scale atmospheric processes and weather conditions establish the potential for severe bushfires, and control the behaviour of individual fires once they begin (Lindesay 2003).

Fire intensity is also influenced by climate and weather variables such as those with relatively long temporal scales such as drought, those with moderate temporal scales such as the diurnal cycle and variables with short temporal scales such as wind (Gill & Bradstock 2003). Many Australian fires that caused the greatest number of deaths, injuries and economic impacts occurred during droughts associated with El Niño events (Hennessy et al. 2005).

By influencing factors such as topography, micro-climates, soil moisture and species composition, climate also plays a role in the types of ecosystems, land uses and human activities in a landscape.

However, Gill and Bradstock (2003) postulated that a fire regime is an emergent property of abiotic and biotic factors, not just climate, vegetation, fuels, terrain, ignition sources or herbivore load alone. We argue similarly that bushfire vulnerability depends on interactions between all those factors, plus the underlying vulnerabilities and resilience of the social and ecological systems affected.

Vulnerability and resilience

Bushfire vulnerability and resilience[1] in both social and ecological systems occur in dynamic and complex ways – spatially, temporally, ecologically and socially and in synergies between these. At basic levels, it can be a consequence of an asset's position in the landscape – the most obvious and frequently discussed example is the urban–rural interface – and/or a function of fire intensity.

Fire suppression becomes increasingly difficult as fire intensity escalates. The capability of suppression action has an upper limit of about 3500 kW/m (3500 kilowatts of energy per linear metre of fire front) in forest fuels. However, bushfire intensities can exceed 100 000 kW/m on 'blow-up' days, such as Ash Wednesday (Incoll 1994). Bradstock and Gill (2001) found that the probability of a house being destroyed by bushfire in the Sydney region increased to 100% when the FFDI was over 40 – the upper end of very high fire danger.

Bushfire vulnerability, however, is more than a consequence of fire intensity, landscape position or whether those at risk simply 'know what to do during a bushfire', as vital as that is. Having a properly prepared home and property is only part of the story and only part of what makes us vulnerable or resilient. Our actions during a fire play one of the greatest roles in bushfire survival. Research consistently shows that survival of people and houses during bushfire depends on a suite of factors that include resident preparedness and behaviour, building construction and property preparedness. Yet it has been long known that implementation of recommended property and personal preparedness activities is influenced by a raft of factors, not just knowledge of risk and risk management options, or even recent bushfire experience.

Economic capacity is a key factor in where people choose to live and it may also have some bearing on their capacity to undertake preparedness actions. Collins (2006) argued that 'rather than being rooted in ignorance, household fire hazard vulnerability is dependent on considerations outside the scope of a biophysical framework'. These include freedom of location choice, economic factors and lifestyle issues. McGee (2005) found that in Edmonton, Canada, most preparation measures were completed for reasons other than fire mitigation. Economic capacity is also a factor in the level of insurance a household takes out. In 1994, concern was expressed at the number of properties damaged or destroyed in the NSW bushfires that were underinsured or not insured at all, and that 'compensation payments to inadequately insured persons should not be seen as a precedent for future disasters' (Ellis et. al. 2004). It would be concerning if those least able to afford insurance are also those with the least capacity to implement appropriate preparedness and response activities.

Bushfire resilience is likely to be affected by health. Together with the danger of physical injury, climatic hazards (such as bushfire) increase exposure to disease pathogens and/or their vectors, exposure to airborne allergens and chemical residuals, and psychosocial and mental health outcomes associated with loss, disruption and displacement (Few 2007). Water shortages, smoke inhalation, burning injuries, loss of income, regional food shortages and psychosocial responses are all potential health impacts of bushfire (Few 2007). Mental and physical health is also likely to influence preparedness and response. Communities such as those involved in the 2003 north-east Victorian fires spent weeks clouded in smoke, waiting for the fires to arrive. Some never saw actual fire, but they had to remain prepared and vigilant, with the extended stress of wondering if and when the fires would arrive.

Just as capacity to implement appropriate planning, preparedness and response activities influences bushfire vulnerability and resilience, so does capacity to recover from the passage of bushfire. Economic rationalism and globalisation expose communities and individuals to market fluctuations. Combined with loss of crops, livestock or other agricultural values as a result of changes in weather such as drought, such a phenomenon has been referred to as 'double-exposure' (Schipper & Pelling 2006). If affected people must also face a bushfire, it may present a threshold beyond which they have little, if any, capacity for response or recovery. More than extending the double-exposure concept, this scenario is intended to highlight the complexities of potential impacts of climate change on community bushfire resilience.

Rural decline can also affect a community's bushfire vulnerability. Population movement from non-urban to metropolitan areas is driven by government withdrawal of services, by globalisation and economic rationalist agendas, from areas with shrinking demand for those services (McKenzie 1999). With fewer people and possibly fewer relevant skills, communities may have less capacity to cope with the passage of a bushfire, or to recover from it. Certainly the current decrease in volunteer numbers is driven in part by globalisation and deregulation of the economy, and technological innovation. These factors make it more difficult for community members to volunteer, regardless of their motivation to do so, in turn making it difficult for Australian volunteer rural fire agencies to meet their community protection responsibilities (McLennan & Birch 2005).

Community resilience is also influenced by ecosystem resilience. Our ecosystems are vulnerable because of pressures such as habitat fragmentation and weed invasion, along with markedly altered fire regimes. Quests to suppress fire have obscured the interdependency of fire and Australian living systems (Main 2004). Regimes are often out of kilter with ecological needs and re-introducing appropriate fire to landscapes is difficult because of myriad land management and social issues. Managing bushfire risk in the urban–rural interface, for example, involves a large number of policy problems and issues, including pest and weed management, water quality, land-use planning, lifestyle choices, visual amenity, habitat maintenance and native vegetation rehabilitation and/or management (Tarrant 2006). Although we are attempting to find a balance, many human and ecological communities are vulnerable (not resilient) because of altered fire regimes.

Climate change

The International Panel on Climate Change (IPCC) Fourth Assessment report (2007) stated that the evidence for climate warming is unequivocal. Even if efforts to reduce emissions are successful, there is a certain amount of change locked into the system: 'Both past and future anthropogenic CO_2 emissions will continue to contribute to warming for more than a millennium and it is very likely that hot extremes and heatwaves will continue to become more frequent' (IPCC 2007a).

More intense and longer droughts are likely, more frequent hot days and nights are virtually certain and the observed increasing frequency of heatwaves is very likely to continue (IPCC 2007a). The most vulnerable industries, settlement and societies include those in areas prone to extreme weather events, especially where rapid urbanisation is occurring. Where extreme weather events become more intense and/or more frequent, the economic and social costs of those events will increase, and these increases will be substantial in the areas most directly affected (IPCC 2007b).

Since 1950, Australia's average temperature has increased by 0.85°C, rainfall has decreased in the south-east, droughts have become hotter and the number of extremely hot days has risen (Hennessy et al. 2006). 2005 was Australia's warmest year on record (BoM 2006). The dry conditions in southern and eastern Australia in 2006 continued the long-term rainfall deficiencies in many regions, some of which have lasted more than five years. Aspects of this multi-year drought are highly unusual and unprecedented in many areas (BoM 2007).

'Even the most stringent mitigation efforts cannot avoid further impacts of climate change in the next few decades, which makes adaptation essential, particularly in addressing near-term impacts' (IPCC 2007b), such as the projected increase in fire-risk weather.

Climate change and bushfires

A growing number of modelling studies project that, due to inevitable changes in climate, the risk of fire weather across fire-prone areas of our planet will increase, including in Australia and New Zealand (see Table 16.1 for a selection of more recent studies and key findings). 'Climate change will exacerbate high temperatures, increased severity and frequency of droughts, and extreme weather events, and this has the potential to affect the intensity, return periods and spread of wildfire' (Hennessy et. al. 2005).

Drier and windier conditions are liable to result in faster fire spread and greater areas burned (Pearce et al. 2005). Such impacts are likely to lead to longer fire seasons and fire danger periods, including an increased number of total fire ban days. A similar picture is painted for regions such as North America, Russia and Europe. In the future, with a warmer climate, we expect more severe fire weather, more area burned, more ignitions and a longer fire season (Flannigan et al., 2005). We may see more lightning-caused fires due to decreased precipitation and increased thunderstorm activity (Goldammer & Price 1998).

Potential climate change impact on extreme events is still being explored, and recent work in Europe suggests that extremes will increase. Extreme events are of particular interest to fire managers as this is typically where the greatest social and ecological bushfire damage occurs. Uncertainties in downscaling global models to regional scales, however, make it difficult to project how the changes may manifest.

In summary, we are likely to see more very high to extreme fire danger days, and more high-intensity, fast and large fires. Fire seasons and fire danger periods are likely to be extended, including an increased number of total fire ban days. Vegetation may have increased flammability and there may be greater fuel availability. Altered fire regimes, in particular increased fire frequencies, will alter ecosystems and impact biodiversity. Greater areas of social, economic and environmental assets are likely to be at risk for extended periods.

Potential community bushfire resilience impacts

Climate change will affect all aspects of our environment, society and economy. However, because of the complex processes and interrelationships and our insufficient understanding of them (Schneider 2004), local manifestations of impacts are uncertain. The IPCC (2007b) stated

Table 16.1 Some recent studies and their projected impacts of climate change on bushfire

Report	Region/ country	Projected impacts
IPCC (2007b)	Global	Resilience of many ecosystems likely to be exceeded by unprecedented combination of climate change, associated disturbances (e.g. bushfire) and other global change drivers Increased deaths, disease and injury due to heatwaves, fires, storms and droughts, particularly in people/fauna/flora with low adaptive capacity
Pitman et al. (2007)	Australia	Increase in regional fire risk (FFDI & GFDI) Single-point 25% increased probability of extreme risk by 2050 (low- or high-emission scenarios) 45% by 2100 if low emissions, to at least 50–100% in Brisbane and Sydney under high-emission scenarios
Hennessy et al. (2005)	South-east Australia	Increase in frequency of very high to extreme fire danger by 2020 and 2050 High-emission scenarios showed a greater increase in risk
Pearce et al. (2005)	New Zealand	More days of very high and extreme forest fire danger Several locations where total number of days of very high/ extreme forest fire danger projected to increase >20 days (>50%) Increase of about 10% in mean westerly wind speed by 2050, and 60% increase by 2080s Increased initial spread index (ISI), resulting in increased [fuel] drying
Moriondo et al. (2006)	European Mediterranean countries	A general increase in fire risk, due mainly to increases in the number of years with fire risk, length of fire season and number of extreme events High-emission scenarios showed a greater increase in risk
Flannigan et al. (2005)	Siberia, Canada, Alaska	Increases in seasonal severity, averaging nearly 50% increase Increased area burned, more than 74–118% increase in area burned by end of 21st century under 3 x CO_2 scenario
Brown et al. (2004) & Kitzberger et al. (2007)	Western US	Increases in the numbers of high fire danger days at least until 2089 Increase in number of bushfires in coming decades

that Australia is likely to have water security problems, significant loss of biodiversity by 2020 and declining agricultural and forestry production by 2030 over much of the south and east due to increased drought and fire. Parts of eastern New Zealand will suffer likewise The report considered that Australian society and economy have substantial adaptive capacity, but there are considerable constraints to implementation and major challenges from extreme events, and that Australia's natural systems have limited adaptive capacity. Unless our policies are robust or adaptive enough to deal with uncertainty, the issues and pressures including vulnerabilities that already contribute to the complexity of bushfire management are likely to worsen. In turn, bushfires are likely to exacerbate some of the adverse effects of climate change. Some of these potentials are outlined below, such as fire suppression capacity, and the economic impact of fires through damage to water supply catchments, additional greenhouse gas emissions and contribution to rural decline.

More very high and extreme fire danger days would raise the chance of residential loss. As mentioned earlier, the chance of house loss is greatest when the FFDI is over 40. The high degree of damage in the 2003 Canberra fires was partly related to the unusual fire severity generated by extreme weather conditions – particularly strong winds combined with drought conditions (Blanchi et al. 2006). Clearly there were other contributing factors to the degree of loss in those fires, such as preparedness and actions on the day. However, increases in fire weather conditions that mimic those in Canberra are liable to magnify the probability of house loss. The combination of heightened risk of house loss and more houses at risk at any one time, added to no or insufficient insurance, holds financial implications for government. There is also potential for insurance erosion, leaving affected communities to deal with event hardships and unable to insure properties because of rising or unavailable premiums (Coleman n.d.).

Increased fire frequency as a result of changing climate is likely to alter ecosystem dynamics, changing vegetation types and biomass fuel loads and types (Cary 2002; Mouillot et al. 2002). How this may manifest is not clear. Some authors argue that an increase in drought frequency could significantly decrease photosynthesis and increase respiration, reducing vegetation productivity and resistance to insects and diseases, resulting in increased stress-induced mortality with subsequent fuel build-up (Goldammer & Price 1998). Others, such as Dyer et al. (2002) also suggest that there may be increased fuel loads resulting from a different mechanism – increased growth associated with the carbon fertilisation effect – but that the increase could be offset to some extent by increased levels of drought stress.

Increases in fire danger imply potentially higher fire intensities, longer fire seasons and more total fire ban days. All this will place increasing pressure on fire suppression resources, as well as fire management costs. The physical realities of fire suppression limitations under higher fire intensities, or a reduced availability of trained personnel to manage prescribed fires or bushfires, may diminish management capacity. Australia's fire suppression resources are often stretched during 'campaign' fires which may run for weeks. If there are more campaign fires, there will be increased personal and employer cost for volunteers along with reliance or pressure on the goodwill of employers and volunteers (Hennessy et al. 2005). More total fire ban days will draw volunteer firefighters away from their work to staff fire stations for rapid response. It may also affect tourism, if people avoid areas they consider high risk on those days.

Bushfire resilience and fire management could be challenged in other ways. Less rain may mean less water is available for suppression by agencies or for active house defence by individual householders. Compounding this, catchments may face an increased risk of post-fire floods if their protective forests burn, affecting water quality and yield. Conversely, projections suggest that when it does rain the events will be more extreme. If such events occur post-fire, as happened in Victoria in 2007, recovery capacity may be diminished and general resilience affected.

Climate change has potential to alter rural landscapes and there is a growing body of work that indicates global warming is likely to increase rural poverty. Research in the US shows that repeated bushfires worsen poverty in some rural communities (Lynn 2003). It is not clear whether this would happen in Australia, but it is clear that climate change is likely to most dramatically affect regions already stressed by high temperatures and low precipitation. Analysis suggests that if climates in such areas become more hostile to farming, there will be a negative impact on rural incomes (Mendelsohn et al. 2007). This may in turn intensify the exodus from much of rural Australia, with implications for local economies and the already-decreasing volunteer base. We discussed an example earlier of agricultural vulnerability, where farmers exposed to market fluctuations may not be able to recover from a bushfire if they have also lost agricultural values to drought. As climate change extends the length and/or severity of droughts and the risk of bushfires, the vulnerability of these people increases. This rather gloomy picture is counterbalanced by other areas, especially those near the coast, having new residents such as people seeking a more relaxed lifestyle ('sea-changers' and 'tree-changers') or lower-cost housing than that available in Australia's metropolitan areas, even though this may contribute to the growing urban–rural interface.

Essential services may be increasingly vulnerable to the rising bushfire risk, as well as to other impacts of climate change. Along with water quality and quantity, services such as electricity, gas and oil, fixed-line telecom networks, rail, bridges, airports, urban facilities and building and structures generally face enhanced risk (Victorian Government 2007).

More subtle impacts may include challenges to risk communications, and a need to reconsider the nuances and sophistication of planning and preparedness advice. There is already a growing demand for information before, during and after fires from communities, media and politicians. Some agencies have found that 'information-flow' programs can provide timely, accurate and appropriate information. Any amplification in the size and numbers of bushfires will increase expectations and the need to continue providing information. There will of course be economic and personnel costs in doing this. Translating more sophisticated concepts such as 'stay or go' into practice is complex, and if days of extreme fire danger are frequent – and are likely to increase with climate change – the approach may become impractical (Handmer & Tibbits 2005). This means climate change impacts on bushfire risk will not diminish 'the tension between the ideology and practical need of asking those at risk to take more responsibility for managing their own risks' (Handmer & Tibbits 2005). Simple advice regarding preferred or less-flammable plants, for example, is likely to be increasingly irrelevant (and at worst misleading) as there is greater potential for extreme fire intensities in which any vegetation will burn, along with heightened pressures on Australia's already vulnerable biodiversity. In a national inquiry, Ellis et al. (2004) wrote that 'there is strong evidence that well-informed and well-prepared communities with realistic expectations of both the likely impacts of bushfire and suppression capabilities of fire services can minimise the impacts of bushfires'. Climate change has the potential to alter what is realistic in terms of bushfire impacts, preparedness and suppression capabilities in a changing climate, and thereby what is communicated (and how).

There is also potential for increasingly stressed populations to put further burdens on ecosystems and services, including strategies to mitigate and adapt to climate change. Climate change will do more than alter ecosystem fuel loads. It is likely to be the greatest threat to biodiversity, and many of the most severe impacts will probably stem from interactions between threats rather than climate change acting in isolation (Thomas et al. 2004). This remains a challenge for ecologically sustainable fire management objectives and communication with local communities.

When bushfires occur, debates, tensions, vested-interest and politically motivated arguments surrounding fire management and associated land management issues all raise their heads. Issues include the growing commercial use of forests or native vegetation protection for carbon sequestration, the concomitant need to protect such sinks from destructive fires, and the increasing need to protect remaining vegetation for biodiversity and ecosystem services (particularly with respect to water in Australia). Blame is placed on the hazard rather than on the conditions of vulnerability resulting from, for example, approaches to governance or marginalisation of sectors of society (Schipper & Pelling 2006), the lack of appropriate funding and resources for land management, underlying vulnerabilities such as those created by economic policy and marginalisation of the land, poor private land management, or inappropriate land use planning. If there are more large fires and we do not begin to address underlying vulnerabilities, these debates and associated inquiry costs will surely escalate.

Finally, there appears to be an assumption in some sectors that climate change will be gradual and that there will be sufficient time to develop adaptation strategies based on past experience with climate variability. This may be true to some extent, but significant change may have already occurred and approaches that assume a near-term return to previous conditions may not work in the face of longer-term change. Moreover, climate change may hold surprises which will not be gradual (Kerr 2007). Unfortunately, most climate change assessments rarely consider low-probability but high-consequence extreme events – it is not even clear that all climate change surprises are actually low-probability. They are just very uncertain at this point, given that the state of knowledge is still evolving (Schneider 2004).

These changes may be well outside our experience in climate variability.

Conclusion

Climate change forces us to think holistically. Yet climate change itself is sometimes pitted against other issues such as bushfires and a choice in dealing with one or the other is implied, as if the two are not inextricably linked. Climate change adaptation requires a holistic long-term perspective that considers not only the risks, opportunities and limitations posed by current and future climate conditions, but also societal changes (OECD 2005). Policies to reduce disaster and climate change risks make sustainable development a core requirement, and place more emphasis on the need to identify and support generic adaptive capacity and move towards the mission of the International Strategy for Disaster Reduction which aims at building disaster-resilient communities (Schipper & Pelling 2006; Haque & Burton 2005).

The concept of disaster-resilient communities is unlikely to be foreign to most fire agencies. Most agencies aim to support communities' ability to cope with bushfire, and have established policy frameworks aimed at achieving such goals. For example, the Country Fire Authority's 2007–08 corporate vision is 'a state where communities partner with CFA to be 'safe, self-reliant, resilient and strong'. However, as the climate changes, the factors that make our communities vulnerable or resilient are likely to change. Supporting social and ecological adaptation or resilience to an increasing (and possibly changing) bushfire risk is a key issue for fire management agencies.

Climate change adaptation represents formidable challenges to government and science, and ultimately to the sustainability of society and the environment on which it depends (Adger et al. 2005). Part of the challenge lies in the fact that we do not know precisely how climate change will affect pertinent variables, limiting our ability to predict consequential effects (Haque & Burton 2005). This in turn limits our ability to develop anticipatory approaches that assume a predictable future.

Policies, trials and management will seem to operate effectively as long as the system remains within known stability domains (Clark et al. 1979). For fire management, climate change is likely to mean that stability domains such as fire season length, opportunities for ecological burning, interactions between vegetation and climate and land-use are no longer well-known and are likely to change for some time. This challenges the notion that the past is a sound guide to the future for bushfire risk analysis and management (Handmer 2003). The need for climate change adaptation, including an increasing bushfire risk, suggests not only that fire management policies and programs may need evaluation in a new light, but that polices that interact with fire management are also implicated. Not least of these is land use planning that creates the ever-expanding urban–rural interface, and that can reduce or contribute to bushfire vulnerability.

Agencies will need to explore what is realistic in terms of bushfire impacts, preparedness and management in a changing climate, and how this will inform their planning, policies and dialogue with communities. Both ecological and social recovery will need to contribute to improved resilience to bushfires and to a changing climate more generally, rather than re-establishing vulnerabilities.

In 2003 Gill and Bradstock argued that given the inherently variable and stochastic nature of fire regimes and their ecological effects (likely to include unpredictable large-scale events that may be impossible to control), management systems and policies must be able to deal with variability and uncertainty. We suggest that there are likely to be more unpredictable large-scale events, and that there is a need for robust yet flexible management and policies.

The resilience perspective shifts policies from aspiring to control change in systems that are assumed to be predictable, to managing the capacity of social–ecological systems to cope with, adapt to and influence change (Smit & Wandel 2006). To support social and ecological resilience or adaptive capacity to an increasing bushfire risk, we need to better understand: the synergies and interrelationships between climate, ecology and society; existing vulnerabilities and ways we can help address them; whether or not our policies are effective; and whether our policies are robust enough to remain effective as impacts of our changing climate manifest.

Uncertainties about the extent and timing of climate change and its synergies with other processes are not an excuse for inaction. As natural hazard disaster managers, fire agencies are perhaps among the most adept practitioners of risk management, including dealing with uncertainties. Using research to build upon that experience and knowledge, we can move along the path of integrated and ecologically sustainable fire management, which will support community adaptation to climate change. Because our climate is changing, now more than ever, if we do not learn to live with fire we will become increasingly vulnerable to it.

Acknowledgments

We thank the Country Fire Authority of Victoria, particularly Noreen Krusel, the Department of Sustainability and Environment's Fire and Emergency Management and Greenhouse Policy Units, particularly Liam Fogarty and Rod Anderson, and the Bushfire CRC for supporting Karyn's research on fire management policy, climate change and social–ecological bushfire resilience.

Endnote

1   We acknowledge the debates surrounding the many definitions of 'vulnerability' and 'resilience'. For simplicity, we use resilience as the opposite of vulnerability. The question is discussed further in Adger (2006) and Fussel (2007) as examples.

References

2: Community perceptions of bushfire risk

Anderson-Berry LJ (2003). Community vulnerability to tropical cyclones: Cairns, 1996–2000. *Natural Hazards* 30: 209–232.

Balcombe L (2007). Perceptions of preparedness for bushfire: a case study of Tamborine Mountain. MSc thesis, James Cook University, Townsville.

Beringer J (2000). Community fire safety at the urban/rural interface: the bushfire risk. *Fire Safety Journal* 35: 1–23.

Best J (1987). Rhetoric in claims-making. *Social Problems* 34(2): 101–121.

Burby R (2001). Involving citizens in hazard mitigation planning: making the right choices. *Australian Journal of Emergency Management* 16(3): 45–51.

Bushnell S, Cottrell A, Spillman M & Lowe D (2006). *Thuringowa bushfire case study: technical report.* James Cook University, Townsville.

Byrd TL & VanDerslice J (1996). *Perception of environmental risk in three El Paso communities.* Proceedings of the HSRC/WERC Joint Conference on the Environment (1996), Great Plains/Rocky Mountain Hazardous Substance Research Centre. Accessed from http://www.p2pays.org/ref/02/01962.pdf on 20/11/06.

Collins TW (2005). Households, forests and fire hazard vulnerability in the American West: a case study of a Californian community. *Environmental Hazards* 6: 23–37.

Cottrell A (2005). Sometimes it's a big ask, but sometimes it's a big outcome: community participation in flood mitigation. *Australian Journal of Emergency Management* 20(3): 27–32.

Covello VT & Johnson BB (1987). The social and cultural construction of risk: issues, methods and case studies. In *The social and cultural construction of risk* (eds BB Johnson & VT Covello). D. Reidel Publishing: Dordrecht.

Denzin NK & Lincoln YS (eds) (1994). *Handbook of qualitative research.* Sage: Thousand Oaks.

Flint CG & Luloff AE (2005). Natural resource-based communities, risk and disaster: an intersection of theories. *Society and Natural Resources* 18: 399–412.

Gilbert JB (2004). *The bushfire risk: community awareness and perception on the rural–urban fringe of Melbourne.* Unit dissertation, University College, London. Accessed from www.nillumbikratepayers.asn.au/the_bushfire_risk.html on 04/08/05.

Gough J (2000). Perceptions of risk from natural hazards in two remote New Zealand communities. *Australasian Journal of Disaster and Trauma Studies.* Accessed from http://www.massey.ac.nz/~trauma/issues/2000-2/gough.htm on 22/01/07.

Hannigan AJ (1995). *Environmental sociology: a social constructionist perspective.* Routledge: London.

Hay I (ed.) (2005). *Qualitative research methods in human geography,* 2nd edn. Oxford University Press: Oxford.

Hoffman S & Oliver-Smith A (2002). *Catastrophe & culture: the anthropology of disaster.* School of American Research Press: Santa Fe.

Holstein JA & Miller G (eds) (1993). *Reconsidering social constructionism: debates in social problems theory*. Alsine de Gruyter: New York.

Johnston DM, Bebbington MS, Lai C-D, Houghton BF & Paton D (1999). Volcanic hazard perceptions: comparative shifts in knowledge and risk. *Disaster Prevention and Management* 8(2): 118–126.

Kreps GA & Drabek TE (1996). Disasters are non-routine social problems. *International Journal of Mass Emergencies and Disasters* 14(2): 129–153.

Kumagai Y, Carroll MS & Cohn P (2004). Coping with interface wildfire as a human event: lessons from the disaster/hazards literature. *Journal of Forestry* 102(6): 28–32.

Lupton D & Tulloch J (2002). Risk is part of your life: risk epistemologies among a group of Australians. *Sociology* 36(2): 317–334.

Lupton D (1999). *Risk*. Routledge: London.

McCaffrey S (2004). Thinking of wildfire as a natural hazard. *Society and Natural Resources* 17: 509–516.

McGee TK & Russell S (2003). 'It's just a natural way of life…': an investigation of wildfire preparedness in rural Australia. *Environmental Hazards* 5: 1–12.

Montz BE (1993). Hazard area disclosure in New Zealand: the impacts on residential property values in two communities. *Applied Geography* 13: 225–242.

Neuman WL (1997). *Social research methods: qualitative and quantitative approaches*, 3rd edn. Allyn & Bacon: Boston.

Odgers P (2002). Bushfire threat and community perception. In *Proceedings of the April 2002 symposium on fire in ecosystems of south-west Western Australia: impacts and management*. Volume 2: Community perspectives about fire, pp. 58–63. Organised by the Department of Conservation and Land Management. Accessed from http://science.calm.wa.gov.au/articles/2003-06-04/fire_proceedings_day3.pdf on 21/10/05.

Odgers P & Rhodes A (2002). *Community response to the New South Wales bushfires 2001–2002*. Australasian Fire Authorities Council: Sydney.

Oliver-Smith A & Hoffman SM (eds) (1999). *The angry earth: disaster in anthropological perspective*. Routledge: New York.

Renn O (2003). Social amplification of risk in participation: two case studies. In *The social amplification of risk* (eds N Pidgeon, RE Kasperson & P Slovic), pp. 374–401. Cambridge University Press: Cambridge.

Rural Fire Service (2002a). *Bushfire risk analysis for Beaudesert Shire*. Unpublished report produced by Rural Fire Service, Queensland Fire and Rescue Service.

Rural Fire Service (2002b). *Bushfire risk analysis for Thuringowa Shire*. Unpublished report produced by Rural Fire Service, Queensland Fire and Rescue Service.

Sjoberg L (1999). Consequences of perceived risk: demand for mitigation. *Journal of Risk Research* 2(2): 129–149.

Spector M & Kitsuse JI (1987). *Constructing social problems*. Aldine de Gruyter: New York.

Stallings RA (1995). *Promoting risk: constructing the earthquake threat*. Aldine de Gruyter: New York.

Watson G (2001). *Tamborine Mountain escarpment management strategy*. Unpublished report to Steering Committee, Tamborine Mountain Escarpment Management Strategy Project, led by Beaudesert Shire Council, Beaudesert.

Winter G & Fried JS (2000). Homeowner perspectives on fire hazard, responsibility, and management strategies at the wildland–urban interface. *Society and Natural Resources* 13: 33–49.

3: Resilience at the urban interface

Bradstock RA, Gill AM, Kenny BJ & Scott J (1998). Bushfire risk at the urban interface estimated from historical weather records: consequences for the use of prescribed fire in the Sydney region of south-eastern Australia. *Journal of Environmental Management* 52: 259–271.

Bullock JA & Haddow GD (2003). *Introduction to emergency management.* Butterworth-Heinemann: Burlington.

Chess C, Salomone KL & Hance BJ (1995). Improving risk communication in government: research priorities. *Risk Analysis* 15: 127–135.

Comfort L (1999). *Shared risk: complex systems in seismic response.* Pergamon: New York.

Cottrell A (2005). Sometimes it's a big ask, but sometimes it's a big outcome: community participation in flood mitigation. *Australian Journal of Emergency Management* 20(3): 27–32.

Gledhill JH (2003). Community self-reliance during bushfires: the case for staying at home. *Third International Wildland Fire Conference and Exhibition,* Sydney, 3–6 October 2003.

Handmer J & Tibbits A (2005). Is staying at home the safest option during bushfires? Historical evidence for an Australian approach. *Global Environmental Change Part B: Environmental Hazards* 6(2): 81–91.

Haynes K, Barclay J & Pidgeon N (in press). Whose reality counts? Factors affecting the perception of volcanic risk. *Journal of Volcanology and Geothermal Research.*

Lowe T & Haynes K (in prep.). Prepare, stay and defend: the Community Fire Unit approach for promoting resilience.

Nilson D (1995). Disaster beliefs and ideological orientation. *Journal of Contingencies and Crisis Management,* 3(1): 12–17.

Pearce L (2003). Disaster management and community planning, and public participation: how to achieve sustainable hazard mitigation. *Natural Hazards* 28: 211–228.

Pitman AJ, Narisma GT & McAneney J (in press). The impact of climate change on Australian bush fire risk. Submitted to *Climatic Change.*

Putnam R (2000). *Bowling alone: the collapse and revival of American community.* Simon & Schuster: New York.

Rohde MS (2002). Command decisions during catastrophic urban-interface wildfire: a case study of the 1993 Orange County Laguna fire. Unpublished Masters thesis, California State University, Long Beach. Accessed on (http://www.wildfirelessons.net/documents/Case_Study_Laguna_Fire_11_93_Rohde02.pdf).

Slovic P (2000). *The Perception of risk.* Earthscan: London.

Wilson AAG & Ferguson IS (1984). Fight or flee? A case study of the Mount Macedon bushfire. *Australian Forestry* 47(4): 230–236.

Wisner B, Blaikie P, Cannon T & Davies I (2004). *At risk: natural hazards, peoples vulnerability and disasters.* 2nd edn. Routledge: London.

4: The concept of local knowledge in rural Australian fire management

Berkes F (1993). Traditional ecological knowledge in perspective. In *Traditional ecological knowledge: concepts and cases* (ed JT Inglis), pp. 1–9. Canadian Museum of Nature, Ottawa, Ontario.

COAG (Council of Australian Governments) (2004). Report of the National Inquiry on Bushfire Mitigation and Management. COAG, Canberra.

Curtis A & Cooke P (2006). *Landcare groups in Victoria: after twenty years.* Report to the Federal Dept of Agriculture, Fisheries and Forestry, National Landcare Program, Monitoring and Evaluation Project.

Esplin B, Gill AM & Enright N (2003). Report of the inquiry into the 2002–2003 Victorian bushfires. State Government of Victoria, Melbourne.

Government of South Australia (2005). *Collaboration is the key. Lessons from the South Australian Government's recovery operation, Lower Eyre Peninsula bushfire, January 2005.* Government of South Australia, Adelaide.

Graeber D (2006). Beyond power/knowledge: an exploration of the relation of power, ignorance and stupidity. Malinowski Memorial Lecture 2006, London School of Economics. Accessed from http://ww.lse.ac.uk/collections/LSEPublicLecturesAndEvents/pdf/20060525-Graeber.pdf

Howitt R (2003). Local and non-specialist participation in impact assessment. In *Strategic management of environmental and socio-economic issues: a handbook* (eds C-Q Liu, Z Zhao, T Xiao & J Guha),pp. 27–36. Guizhou Science and Technology Publishing House: Guiyang, China. Accessed from http://www.es.mq.edu.au/~rhowitt/SIA_00E9.HTM.

Mahiri I (1998). Comparing transect walks with experts and local people. *PLA Notes* 31: 4–8.

Manne RA (2007). Turkish tales: Gallipoli and the Armenian genocide. *The Monthly* February: 20–28.

McLeod R (2003). Inquiry into the operational response to the January 2003 bushfires in the ACT. ACT Government, Canberra.

Parliament of the Commonwealth of Australia (2003). *A nation charred.* Nairn Report. Parliament, Canberra.

Pearce L (2003). Disaster management and community planning, and public participation: how to achieve sustainable hazard mitigation. *Natural Hazards* 28: 211–228.

Robbins P (2004). The politics of barstool biology: environmental knowledge and power in greater Northern Yellowstone. *Geoforum* 37: 185–199.

Robson C (2002). *Real world research.* Blackwell: Oxford.

Stretton LEB (1939). *Report of the royal commission to inquire into the causes of and measures taken to prevent the bush fires of January, 1939, and to protect life and property and the measures to be taken to prevent bush fires in Victoria and to protect life and property in the event for future bush fires.* Government Printer, Melbourne.

Yli-Pelkonen V & Kohl J (2005). The role of local ecological knowledge in sustainable urban planning: perspectives from Finland. *Sustainability: Science, Practice and Policy* 1(1): 2.

Zylstra P (2006). Fire history of the Australian Alps: prehistory to 2003. Australian Alps Liaison Committee, NSW National Parks and Wildlife Service. Accessed from http://www.australianalps.deh.gov.au/publications/fire/history.html.

5: Social contexts of responses to bushfire threat

Beatson R & McLennan J (2005). Australia's women volunteer fire fighters: a literature review and research agenda. *Australian Journal on Volunteering* 10(2): 18–27.

Cheers B (2002). *The community factor: a critical review of the literature.* Centre for Rural and Regional Development, University of South Australia, Whyalla.

Conway ME (1978). Theoretical approaches to the study of roles. In *Role theory. perspectives for health professionals* (eds ME Hardy & ME Conway), pp. 17–28. Appleton/Century-Crofts: Norwalk, CT.

Enarson E & Hearn Morrow B (eds) (1998). *The gendered terrain of disaster: through women's eyes.* Greenwood Publications: Westport, CT.

Enarson E & Meyreles L (2004). International perspectives on gender and disaster: differences and possibilities. *International Journal of Sociology and Social Policy* 24(10/11): 49–92.

Fothergill A (1996). Gender, risk and disaster. *Applied Behavioural Science Review* 7(2): 125–143.

Goodman H, Healey L & Boulet J (2007). Community responses to bushfire. The role and nature of systems of primary sociality. *New Community Quarterly* 5(1): 11–25.

Gould J (2005). Report for the coroner on the Wangary fire. Unpublished.

Issacs W (1999). *Dialogue: the art of thinking together*. Doubleday Currency: New York.

Long J & Honner S (2006). Engaging women in agricultural training. Ag Consulting. Accessed from http://www.farmbis.sa.gov.au/pdfserve/about_farmbis/reports/engaging_women.pdf on 20/11/06.

Position paper on community safety and evacuation during a bushfire. Accessed from http://www.afac.com.au/awsv2/publications/documents/PositionPaperonBushfiresandCommunitySafety.pdf on 20/11/06.

Proudley M (2007). Fires, families and decisions. Presentation at the joint AFAC/Bushfire CRC conference *A Southerly Change*, Tassie Fire Conference, 19–21 September 2007, Hobart.

Rhodes A (2005). Householder preparedness & response in the Wangary bushfire, Lower Eyre Peninsula, South Australia. Unpublished report.

Rogers EM (1995). *The diffusion of innovations*. Free Press: New York.

Segerson C (2002). Evaluation as responsibility, conscience and conviction. In *Exploring evaluation role and identity* (eds K Ryan & T Schwandt), pp. 87–102. Information Age Publishing: Connecticut.

Smith B (2005). Report of independent review of circumstances surrounding Eyre Peninsula bushfire of 10 and 11 January 2005 (Wangary bushfire). Unpublished.

Turner RH (2002). Role theory. In *Handbook of sociological theory* (ed. JH Turner). Kluwer Academic/Plenum Press: New York.

6: Prepare, stay and defend or leave early

Australasian Fire Authorities Council (2005). Position paper on bushfires and community safety. AFAC: Melbourne.

Barrow GJ (1945). A survey of houses affected in the Beaumaris fire, January 14 1944. *Journal of the Council for Scientific and Industrial Research* 18(1).

Brennan P (1998). *Bushfire threat: response in a small community*. Australian Emergency Management Institute: Macedon, Vic.

Country Fire Authority (2004). Living in the bush: bushfire survival plan workbook. Accessed from http://www.cfa.vic.gov.au/residents/living/litb-workbook.htm on 3/3/06.

Country Fire Authority (2006). Essential equipment. Accessed from http://www.cfa.vic.gov.au/residents/living/essentialequipment.htm on 3/3/06.

Deputy State Coroner (2005). Preliminary findings of inquest. Accessed from http://www.courts.sa.gov.au/courts/coroner/findings/findings_2005/wangary_bushfires.htm on 23/12/06.

Gledhill JH (2003). Community self-reliance during bushfires: the case for staying at home. *Third International Wildland Fire Conference and Exhibition*, 3–6 October 2003, Sydney.

Handmer J (2000). Are flood warnings futile? Risk communication in emergencies. *Australasian Journal of Disaster and Trauma Studies* 2.

Handmer J & Tibbits A (2005). Is staying at home the safest option during bushfires? Historical evidence for an Australian approach. *Environmental Hazards* 6(2): 81–91.

Haynes K & Tibbits A (forthcoming). An examination of the trends in civilian bushfire fatalities in Australian over the last 100 years.

Lazarus G & Elley J (1984). A study of the effect of household occupancy during the Ash Wednesday bushfire in Upper Beaconsfield, Victoria, February 1983. Technical Paper No. 3, National Centre for Rural Fire Research.

Leonard J (2003). People and property: a researcher's perspective. In *Australia burning: fire ecology, policy and management issues* (eds G Cary, D Lindenmayer & S Dovers), pp. 103–112. CSIRO Publishing: Melbourne.

Leonard J & McArthur NA (1999). A history of research into building performance in Australian bushfires. In *Proceedings of the Australian Bushfire Conference* (eds I Lunt & DG Green), Charles Sturt University, Albury.

Lowe T, Tibbits A, Handmer J & Whittaker J (in prep.) 'Confidence is key': an assessment of residents' action during the October 2006 bushfires in Hobart.

McArthur AG & Cheney NP (1967). *Report on the Southern Tasmanian bushfires of 7 February 1967*. Forestry Commission Tasmania, Government Printer: Hobart.

Mental Health Research and Evaluation Centre (1985). The Health and Social Impact of the Ash Wednesday bushfires: a survey of the twelve months following the bushfires of February 1983. South Australian Health Commission: Adelaide.

Miller SI, Carter W & Stevens RG (1984). Report of the Bushfire Review Committee on Bushfire disaster preparedness and response in Victoria, Australia, following the Ash Wednesday fires, February 16, 1983. Dept of Police and Emergency Services: Melbourne

Morgan MG, Fischhoff B, Bostrom A & Atman CJ (2002). *Risk communication: a mental models approach*. Cambridge University Press: Cambridge.

Odgers P & Rhodes A (2002). Community response to the NSW bushfires 2001–2002. AFAC: Melbourne

Ramsay GC, McArthur NA & Dowling VP (1987). Preliminary results from an examination of house survival in the 16 February 1983 bushfires in Australia. *Fire and Materials* 11(1): 49–51.

Rhodes A (2005a). *Stay or go: what do people think of the choice?* Paper at Australasian Fire Authorities and Bushfire CRC Conference, 5–7 October 2005, Auckland, New Zealand.

Rhodes A (2005b). *Householder preparedness and response in the Eyre Peninsula bushfire, 2005*. Bushfire Co-operative Research Centre: Melbourne.

Tibbits A & Whittaker J (in press). Stay and defend or leave early: policy problems and experiences during the 2003 Victorian bushfires. *Environmental Hazards*. doi:10.1016/j.enhuhas.2007.08.001

Victorian Coroner (1997). Inquests into the deaths of Jennifer Louise Lindroth, Graham Kingsley Lindroth & Genevieve Erin during a fire at Ferny Creek and four other fires on the Dandenong Ranges on 21 January 1997. Coronial Services Centre: Melbourne

Wilson AAG & Ferguson IS (1984). Fight or flee? A case study of the Mount Macedon bushfire. *Australian Forestry* 47(4): 203–206.

7: Property safety

AFAC (nd). *Guidance for people in vehicles during bshfires* AFAC: Melbourne. Accessed from http://www.afac.com.au/awsv2/publications/documents/Guidance_for_people_in_vehicles_during_bushfires.pdf.

AFAC (2005). *Position paper on bushfires and community safety*. AFAC: Melbourne. Accessed from http://www.afac.com.au/awsv2/publications/documents/PositionPaperonBushfires andCommunitySafety.pdf.

Blanchi R, Leonard JE & Leicester RH (2006a). *Bushfire risk at the rural/urban interface*. Bushfire 2006 conference. Brisbane.

Blanchi R, Lucas C & Finkele K (2006b). *Meteorological conditions and wildfire related house loss in Australia*. AFAC Conference, Melbourne.

Blanchi R, Leonard JE, White N, Bicknell A, Sargent A & Andersson S (2006c). *Research into the performance of water tanks in bushfire*. Bushfire CRC report.

Bowditch PA, Leonard JE, Sargeant AJ & Macindoe L (2005). *Window and glazing exposure to laboratory-simulated bushfires*. Bushfire CRC report.

FESA (2001). *Planning for bushfire protection*. Fire and Emergency Services Authority, Perth.

Handmer J & Tibbits A (2005). Is staying at home the safest option during bushfires? Historical evidence for an Australian approach. *Global Environmental Change B: Environmental Hazards* 6(2): 81–91.

Krusel N & Petris NS (1992). A study of civilian deaths in the 1983 Ash Wednesday bushfires, Victoria, Australia. Occasional Paper No. 1. Country Fire Authority: Victoria.

Leonard JE (2003). *Bushfires in the ACT*. Australian Institute of Building Surveyors 38th Annual State Conference, Canberra.

Leonard JE & Blanchi R (2005). *Investigation of bushfire attack mechanisms involved in house loss in the ACT bushfire 2003*. Bushfire CRC report.

Leonard JE & Bowditch PA (2003). *Findings of studies of houses damaged by bushfire in Australia*. Wildland Fire Conference, Sydney.

Leonard JE, Blanchi R & Bowditch PA (2004). *Bushfire impact from a house's perspective*. Earth Wind and Fire, Bushfire 2004 Conference, Adelaide.

Leonard JE, Blanchi R, White N, Bicknell A, Sargeant A & Reisen F (2005). Research and investigation into the performance of residential boundary fencing systems in bushfires. Bushfire CRC report.

McArthur NA (1997). *A protocol for surveying bushfire building damage*. Australia Bushfire Conference, Darwin.

NSW Rural Fire Service (2006). *Planning bushfire protection*. NSW Rural Fire Service: Sydney.

Ramsay GC, McArthur NA & Dowling VP (1986). *Building survival in bushfires*. 4th Biennial Conference of the Institution of Fire Engineers, Perth.

Ramsay GC, McArthur NA & DowlingVP (1994). *Bushfires: lessons for planners*. Queensland Planners Conference, Brisbane.

Standards Australia (1999). *AS 3959: Construction of buildings in bushfire-prone areas*. Standards Australia: Sydney.

Wilson AAG & Ferguson IS (1984). Fight or flee: a case study of the Mount Macedon bushfire. *Fireman (Operations Supplement)* 156: 1–8.

Wilson AG & Ferguson IS (1986). Predicting the probability of home survival during bushfires. *Journal of Environmental Management* 23: 259–270.

8: Don't get burnt by the law

Commonwealth of Australia (2002). *Final report of the review of the law of negligence* (Ipp Report). 30 September 2002, Canberra.

Dunlop C (2004). Legal issues in emergency management: lessons from the last decade. *Australian Journal of Emergency Management* 19(1): 26–33.

Eburn M (2005). *Emergency law*, 2nd edn. Federation Press: Sydney.

International Association of Wildland Fire (IAWF) (2007). Survey: firefighters may decline duties. *Wildfire* March/April: 6–7.

Karanev K (2001). Assessing the legal liabilities of emergencies. *Australian Journal of Emergency Management* 16(1): 18–22.

Luntz H & Hambly D (2005). *Torts: cases and Commentary*, 5th edn. LexisNexis Butterworths: Sydney.

Timbs v Shoalhaven City Council [2004] NSWCA 81.

Pyrenees Shire Council v Day (1998) 151 ALR 147.

9: Understanding and preventing bushfire arson

Australian Institute of Criminology (2005a). *Age of criminal responsibility*. Crime facts info no. 106. Accessed from http://www.aic.gov.au/publications/cfi/cfi106.html.

Australian Institute of Criminology (2005b). *Causal factors in New South Wales investigated bushfires. Part 1: deliberate fires.* Bushfire Arson Bulletin No. 22. Accessed from http://www.aic.gov.au/publications/bfab/bfab022.html

Baird RA (2006). Pyro-terrorism: the threat of arson-induced forest fires as a future terrorist weapon of mass destruction. *Studies in Conflict and Terrorism* 29: 415–428.

Bryant C (2007). *Understanding bushfire: trends in deliberate vegetation fires in Australia.* Technical and background paper series. Australian Institute of Criminology: Canberra.

Cheney P (1995). *Bushfires: an integral part of Australia's environment.* Australian Bureau of Statistics: Canberra. Accessed from http://www.abs.gov.au/Ausstats/abs@.nsf/Lookup/6C98BB75496A5AD1CA2569DE00267E48.

Dadds MR & Fraser JA (2006). Fire interest, fire setting and psychopathology in Australian children: a normative study. *Australian and New Zealand Journal of Psychiatry* 40(6): 581–586.

Doley R (2003). Pyromania: fact or fiction? *British Journal of Criminology* 43(4): 797–807.

Drabsch T (2003). *Arson.* Briefing Paper 2/2003. Accessed from http://www.parliament.nsw.gov.au/prod/parlment/publications.nsf/Key/ResearchBF022003.

Epps K & Hollin CR (2000). Understanding and treating adolescent firesetters. In *Violent children and adolescents: asking the question why* (ed G Boswell), pp. 36–55. Whurr Publishers: Philadelphia.

Kafry D (1990). Playing with matches: children and fire. In *Fires and human behaviour*, 2nd edn (ed D Canter), pp. 47–62. David Fulton Publishers: London.

Lewis A (1999). *The prevention and control of arson.* Fire Protection Association: Hertfordshire, UK.

Model Criminal Code Officers Committee of the Standing Committee of Attorneys-General (MCCOC), (2001). Model Criminal Code Report. Chapter 4: Damage and computer offences. MCCOC: Australia.

Muller DA & Stebbins A (2007). *Juvenile arson intervention programs in Australia.* Trends and Issues in Crime and Criminal Justice series, no. 35.

National Association of State Fire Marshals (2001). *Juvenile firesetter intervention research project: final report.* Accessed from http://www.firemarshals.org/programs/docs/NASFM%20Final%20Report-Rev101b.pdf.

Productivity Commission (2006). *Report on government services 2006. Part D: Emergency management.* Accessed from http://www.pc.gov.au/gsp/reports/rogs/2006/emergency management/index.html on 16/11/06.

Scholte EM (1999). Factors predicting continued violence into young adulthood. *Journal of Adolescence* 22: 3–20.

Shea P (2002). The lighting of fires in a bushland setting. *Judicial Officers Bulletin* 14(1): 1–4, 8.

Stanley J (2002). Preventing children and young people lighting bushfires in Australia. *Child Abuse Prevention Newsletter* 10(2): 6–11.

Swaffer T & Hollin CR (1995) Adolescent firesetting: why do they say they do it? *Journal of Adolescence* 18: 619–623.

Willis M (2004). *Bushfire arson: a review of the literature.* Research and Public Policy Series, 61. Australian Institute of Criminology: Canberra.

10: The media and fire services

Blong RJ (1985). Public views on disaster response and the news media: some Australian examples. *Proceedings of research workshop on human behaviour in disaster in Australia*, 25–27 April, Australian Counter Disaster College, Macedon, Victoria.

Campbell A (2003). Learning to live with fire. In *Australia burning: fire ecology, policy and management issues* (eds G Cary, D Lindenmayer & S Dovers), pp. 243–247. CSIRO Publishing: Melbourne.

Carson C (2004). The Information Unit: effective information flow to the community at risk during the onset of wildfire in Victoria. *Are we prepared for future challenges?* AFAC Conference, Perth, Western Australia.

Country Fire Authority of Victoria (2000). CFA Media Forum. Unpublished paper. Emergency Management Australia: Macedon, Victoria.

Ellis S, Kanowski P & Whelan R (2004). *National inquiry on bushfire mitigation and management.* Commonwealth of Australia: Canberra. Accessed from http://www.coagbushfireenquiry.gov.au/findings.htm.

Goltz JD (1984). Are the news media responsible for the disaster myths? A content analysis of emergency response imagery. *International Journal of Mass Emergencies and Disasters* 2(3): 345–368.

Klinenberg E (2002). *Heat wave: a social autopsy of disaster in Chicago.* University of Chicago Press: Chicago.

Marshall PD (1994). Panic TV: Australian bushfires and the media future. *Meanjin* 53(3): 537–543.

Quarantelli EL (1989). The social sciences study of disasters and mass media. In *Bad tidings: communication and catastrophe* (eds T Walters, LM Walters & L Wilkins), pp. 1–19. Erlbaum: Hillsdale, NJ.

Quarantelli EL (2002). The role of the mass communication system in natural and technological disasters and possible extrapolation to terrorism situation. *Proceedings of Natural Disasters Roundtable: National Academy of Sciences on Countering Terrorism: Lessons Learned From Natural and Technological Disasters.* Disaster Research Center, Washington DC. Preliminary Paper. Accessed from http://dels.nas.edu/dr/docs/Quarantelli.pdf.

11: Preparing for bushfires

Adams J (1995). *Risk.* UCL Press: London.

Bright AD & Manfredo MJ (1995). The quality of attitudinal information regarding natural resource issues: the role of attitude strength, importance and information. *Society and Natural Resources* 8: 399–414.

Bright AD & Manfredo MJ (1997). The influence of balanced information on attitudes toward natural resource issues. *Society and Natural Resources* 10: 469–483.

Bright AD, Manfredo MJ, Fishbein M & Bath A (1993). Application of the theory of reasoned action to the National Parks Service's controlled burn policy. *Journal of Leisure Research* 25: 263.

Ellis S, Kanowski P & Whelan R (2004). *National inquiry on bushfire mitigation and management.* Commonwealth of Australia: Canberra.

Fried JS, Winter GJ & Gilless JK (1999). Assessing the benefits of reducing fire risk in the wildland urban interface: a contingent valuation approach. *International Journal of Wildland Fire* 9: 9–20.

Kneeshaw K, Vaske JJ, Bright A & Absher JD (2004). Situational influences of acceptable wildland fire management actions. *Society and Natural Resources* 17: 477–489.

Kumagai Y, Bliss JC, Daniels SE & Carroll MS (2004). Research on causal attribution of bushfire: an exploratory multiple-methods approach. *Society and Natural Resources* 17: 113–127.

Lion R, Meertens RM & Bot I (2002). Priorities in information desire about unknown risks. *Risk Analysis* 22: 765–776.

McGee TK & Russell S (2003). 'It's just a natural way of life…': an investigation of wildfire preparedness in rural Australia. *Environmental Hazards* 5: 1–12.

McLeod R (2003). Inquiry into the operational response to the January 2003 Canberra bushfires in the ACT. Department of Urban Services: Canberra.

Paton D (2005). *Community resilience: integrating hazard management and community engagement.* Proceedings of the International Conference on Engaging Communities. Queensland Government/UNESCO: Brisbane.

Paton D (in press). Risk communication and natural hazard mitigation: how trust influences its effectiveness. *International Journal of Global Environmental Issues.*

Paton D & Bishop B (1996). Disasters and communities: promoting psychosocial well-being. In *Psychological aspects of disaster: impact, coping, and intervention* (eds D Paton & N Long). Dunmore Press: Brisbane.

Paton D & Bürgelt PT (2006). Social-ecological vulnerability: factors facilitating co-existence with bushfire hazards. IGU Commission on Population and Vulnerability and Asia Pacific Migration Research Network (APMRN), IGU2006 Conference, 3–7 July, Brisbane.

Paton D, Smith LM & Johnston D (2005). When good intentions turn bad: promoting natural hazard preparedness. *Australian Journal of Emergency Management* 20: 25–30.

Paton D, Kelly G, Bürgelt PT & Doherty M (2006a). Preparing for bushfires: understanding intentions. *Disaster Prevention and Management* 15: 566–575.

Paton D, Kelly G & Doherty M (2006b). Exploring the complexity of social and ecological resilience to hazards. In *Disaster resilience: an integrated approach* (eds D Paton & D Johnston). Charles C Thomas: Springfield, Ill.

Paton D, McClure J & Bürgelt PT (2006c). Natural hazard resilience: the role of individual and household preparedness. In *Disaster resilience: an integrated approach* (eds D Paton & D Johnston). Charles C Thomas: Springfield, Ill.

Turner RH, Nigg JM & Paz DH (1986). *Waiting for disaster: earthquake watch in California.* University of California Press: Los Angeles.

Vogt CA, Winter G & Fried JS (2005). Predicting homeowners' approval of fuel management at the wildland-urban interface using the theory of reasoned action. *Society and Natural Resources* 18: 337–354.

Weinstein ND (1980). Unrealistic optimism about future life events. *Journal of Personality and Social Psychology* 39: 806–820.

Winter G & Fried JS (2000). Homeowner perspectives on fire hazard, responsibility and management strategies and the wildland-urban interface. *Society and Natural Resources* 13: 33–49.

Winter G, Vogt CA & Fried JS (2002). Fuel treatments at the wildland-urban interface: common concerns in diverse regions. *Journal of Forestry* 100: 15–21.

Winter G, Vogt CA & McCaffrey S (2004). Examining social trust in fuels management strategies. *Journal of Forestry* 102: 8–15.

12: The use of program theory

Campbell H (2005). Logic models in support of homeland security strategy development. *Journal of Homeland Security and Emergency Management* 2(2): article 9.

Chen H (1990). *Theory-driven evaluations.* Sage Publications: Newbury Park, CA.

ECONorthwest (2006). NFPA Firewise ArcView: lessons learned. Research project. Fire Protection Research Foundation: MA.

Funnell S (1997). Program logic: an adaptable tool for designing and evaluating programs. *Evaluation News and Comment* July: 5–17.

Funnell S (2000). Developing and using a program theory matrix for program evaluation and performance monitoring. In *Program theory in evaluation: challenges and opportunities, new directions for evaluation* (eds P Rogers, T Hacsi, A Petrosino & T Huebner) 87: 91–101.

Lipsey M & Pollard J (1989). Driving toward theory in program evaluation: more models to choose from. *Evaluation and Program Planning* 12: 317–328.

Patton M (1997). *Utilized-focused evaluation.* Sage Publications: Thousand Oaks, CA.

Pawson R & Tilley N (1997). *Realistic evaluation.* Sage Publications: Thousand Oaks, CA.

Pawson R & Tilley N (2005). Realistic evaluation. In *Encyclopedia of evaluation* (ed. S Mathison). Sage Publications: Thousand Oaks, CA.

Pope J (2006). *Indicators of community strength: a framework and evidence.* Department for Victorian Communities: Melbourne.

13: What should community safety initiatives for bushfire achieve?

Barnes P (2002). Approaches to community safety: risk perception and social meaning. *Australian Journal of Emergency Management* 17(1): 15–23.

Chess C, Salomone KL, Hance BJ & Saville A (1995). Results of a national symposium on risk communication: next steps for government agencies. *Risk Analysis* 15(2): 115–125.

Fellbaum CE (1998). *Wordnet: an electronic lexical database.* Bradford Books: Cambridge, MA.

Hodges A (1999). *Towards community safety.* Paper presented at the Australasian Fire Authorities Conference, Priority Community Safety, Melbourne.

Laverack G & Labonte R (2000). A planning framework for community empowerment goals within health promotion. *Health Policy and Planning* 15(3): 255–262.

Leeuw FL (2003). Reconstructing program theories: methods available and problems to be solved. *American Journal of Evaluation* 24(1): 5–20.

McClintock C (1990). Evaluators as applied theorists. *Evaluation Practice* 11(1): 1–12.

McEntire DA, Fuller C, Johnston CW & Weber R (2002). A comparison of disaster paradigms: the search for a holistic policy guide. *Public Administration Review* 62(3): 267–281.

Novak JD & Gowin DB (1984). *Learning how to learn.* Cambridge University Press: New York.

Pawson R (2006). *Evidence-based policy: a realist perspective.* Sage Publication: London.

Pawson R & Tilley N (1997). *Realistic evaluation.* Sage Publications: Thousand Oaks, CA

Rhodes A & Gilbert J (2007). The use of program theory in the evaluation of bushfire community safety programs. Chapter 12, this volume.

Smith P, Nicholson J & Collett L (1996). Risk management in the fire and emergency services. *Australian Journal of Emergency Management* 11(2): 5–13.

Squires P (1997). *Criminology and the 'community safety' paradigm: safety, power and success and the limits of the local.* Paper presented at the British Criminology Conference, Queens University, Belfast.

Suchman EA (1967). *Evaluative research: principles and practice in public service & social action programs.* Russell Sage Foundation: New York.

Tilley N (2004). Applying theory-driven evaluation to the British Crime Reduction Programme: the theories of the programme and of its evaluations. *Criminal Justice* 4(3): 255–276.

Trochim WMK (1989). An introduction to concept mapping for planning and evaluation. *Evaluation and Program Planning* 12(1): 1–16.

Wishart D (2004). Clustan graphics (Version 6.06). [Cluster analysis]. Clustan: Edinburgh.

WK Kellogg Foundation (2004). *Logic model development guide.* WK Kellogg Foundation: Battle Creek, MI.

Yampolskaya S, Nesman TM, Hernandez M & Koch D (2004). Using concept mapping to develop a logic model and articulate a program theory: a case example. *American Journal of Evaluation* 25(2): 191–207.

14: The value of economics for fire management

Boadman AE, Greenberg DH, Vining AR & Weimer DL (2001). *Cost-benefit analysis: concepts and practice*. Prentice Hall: New Jersey.

Bureau of Transport Economics (2001). *Economic cost of natural disasters in Australia*. Commonwealth of Australia: Canberra.

Country Fire Authority (2005). *Annual Report 2004*. Accessed from www.cfa.vic.gov.au in February 2005.

Department of Sustainability and Environment (2003). Department website www.dse.vic.gov.au (accessed February 2005).

Ellis S, Kanowski P & Whelan R (2004). *National inquiry on bushfire mitigation and management*. Council of Australian Governments: Canberra.

Freeman III AM (2003). *The measurement of environmental and resource values: theory and methods,* 2nd edn. Resources for the Future: Washington DC.

Gangemi M, Martin J, Marton R, Phillips S & Stewart M (2003). A report on the socio-economic impact of business on rural communities and local government in Gippsland and north east Victoria. Centre for Regional and Rural Development, RMIT University: Melbourne.

Glover D & Jessup T (1999). *Indonesia's fire and haze: the cost of catastrophe*. Institute of South East Asian Studies, Singapore & International Development Research Centre: Ottawa.

Handmer J, Read C & Percovich O (2002). *Disaster loss assessment guidelines*. Emergency Management Australia: Canberra.

Hatch JH & Jarret FG (1985). The economics of fire control/suppression. In *The economics of bushfire: the South Australian experience* (eds DT Healey, FG Jarrett & JM McKay), pp. 89–115. Centre for South Australian Economics Studies/Oxford University Press: Melbourne.

Pyne SJ, Andrews PL & Leven RD (1996). *Fire economics: introduction to wildland fire,* 2nd edn. John Wiley & Sons: New York.

Rideout DB & Omi PN (1990). Alternate expression for the economic theory of forest fire management. *Forest Science* 36(3): 614–624.

Russell CS (1970). Losses from natural hazards. *Land Economics* 46: 383–393.

Whittaker J & Mercer D (2004). The Victorian bushfires of 2002–03 and the politics of blame: a discourse analysis. *Australian Geographer* 35(3): 259–287.

15: Save that brigade!

Birch A & McLennan J (2006a). Age and motivation to volunteer. *Fire Australia* Spring: 21–23.

Birch A & McLennan J (2006b). *NSW grain belt community survey 2005: evaluating the community's understanding of the NSW Rural Fire Service*. Bushfire Cooperative Research Centre Volunteerism Project, La Trobe University: Melbourne.

Cyberiad (2005). Exit survey: report to the South Australian Country Fire Service Board. Adelaide.

King S, Bellamy J & Donato-Hunt C (2006). The cost of volunteering: a report on a national survey of emergency management sector volunteers. Anglicare: Sydney.

McLennan J & Birch A (2005). A potential crisis in wildfire emergency response capability? Australia's volunteer firefighters. *Environmental Hazards* 6: 101–107.

McLennan J & Birch A (2006a). *Survey of South Australian Country Fire Service women volunteers*. Bushfire Cooperative Research Centre Volunteerism Project, La Trobe University: Melbourne.

McLennan J & Birch A (2006b). *Survey of ACT Rural Fire Service women volunteers*. Bushfire Cooperative Research Centre Volunteerism Project, La Trobe University: Melbourne.

McLennan J & Birch A (2006c). *Survey of New South Wales employers.* Bushfire Cooperative Research Centre Volunteerism Project, La Trobe University: Melbourne.

Meyer JP, Allen NJ & Topolnytsky L (1998). Commitment in a changing world. *Canadian Psychology* 39: 83–93.

Woodward A & Kallman J (2001). Why volunteers leave: volunteer exit survey in the emergency services. *Australian Journal on Volunteering* 6(2): 91–98.

16: Climate change and community bushfire resilience

Adger WN (2006). Vulnerability. *Global Environmental Change* 16: 268–281.

Adger WN, Arnell NW & Tompkins EL (2005). Adapting to climate change: perspectives across scales. *Global Environmental Change* (editorial) 15: 75–76.

Blanchi R, Leonard J & Maughan D (2006). Towards new information tools for understanding bushfire risk at the urban interface. In *Bushfire 2004: Earth, Wind and Fire*, Conference Proceedings, Adelaide.

BoM (Bureau of Meteorology) (2006). *Annual Australian climate summary 2005 – media release.* Bureau of Meteorology: Melbourne. Accessed from http://www.bom.gov.au/announcements/media_releases/climate/change/20060104.shtml on 31/5/07.

BoM (2007). *Annual Australian climate summary 2006 – media release.* Bureau of Meteorology: Melbourne. Accessed from http://www.bom.gov.au/announcements/media_releases/climate/change/20070103.shtml on 31/5/07.

Bradstock R & Gill M (2001). Living with fire and biodiversity at the urban edge: in search of a sustainable solution to the protection problem in southern Australia. In *Climate change and bushfire incidence* (ed. G Cary). Proceedings of the New South Wales Clean Air Forum 2004.

Brown TJ, Hall BL & Westerling AL (2004). The impact of twenty-first century climate change on wildland fire danger in the western United States: an applications perspective. *Climatic Change* 62: 365–388.

Cary G (2002). Importance of a changing climate for fire regimes in Australia. In *Flammable Australia: fire regimes and biodiversity of a continent* (eds R Bradstock, J Williams & M Gill). Cambridge University Press: Cambridge.

Cary G (2004). Climate change and bushfire incidence. In *Proceedings of the New South Wales Clean Air Forum 2004.*

Clark WC, Jones DD & Holling CS (1979). Lessons from ecological policy design: a case study of ecosystem management. *Ecological Modelling* 7: 1–53.

Coleman T (nd). The impact of climate change on insurance against catastrophes. Insurance Australia Group.

Collins TW (2006). Households, forests, and fire hazard vulnerability in the American West: a case study of a California community. *Environmental Hazards* 6: 23–37.

Dyer R, Jacklyn P, Partridge I, Russell-Smith J & Williams RJ (eds) (2002). *Savanna burning: understanding and using fire in northern Australia.* Tropical Savannas CRC: Darwin.

Ellis S, Kanowski P & Whelan R (2004). *National inquiry on bushfire mitigation and management.* Commonwealth of Australia: Canberra.

Few R (2007). Health and climatic hazards: framing social research on vulnerability, response and adaptation. *Global Environmental Change* 17(2): 281–295.

Flannigan MD, Amiro BD, Logan KA, Stocks BJ & Wotton BM (2005). Forest fires and climate change in the 21st century. *Mitigation and Adaptation Strategies for Global Change* 11: 847–859.

Fussel H-M (2007). Vulnerability: a generally applicable conceptual framework for climate change research. *Global Environmental Change* 17(2): 155–167.

Gill AM & Bradstock R. (2003). Fire regimes and biodiversity: a set of postulates. In *Australia burning: fire ecology, policy and management issues* (eds G Cary, D Lindenmayer & S Dovers). CSIRO Publishing: Melbourne.

Goldammer JG & Price C (1998). Potential impacts of climate change on fire regimes in the tropics based on MAGICC and a GISS GCM_derived lightning model. *Climatic Change* 39: 273–296.

Handmer J (2003). Institutions and bushfires, fragmentation, reliance and ambiguity. In *Australia burning: fire ecology, policy and management issues* (eds G Cary, D Lindenmayer & S Dovers). CSIRO Publishing: Melbourne.

Handmer J & Tibbits A (2005). Is staying at home the safest option during bushfires? Historical evidence for an Australian approach. *Environmental Hazards* 6: 81–91.

Haque CE & Burton I (2005). Adaptation options strategies for hazards and vulnerability mitigation: an international perspective. *Mitigation and Adaptation Strategies for Global Change* 10: 335–353.

Hennessy K, Lucas C, Nicholls N, Bathols J, Suppiah R & Ricketts J (2005). *Climate change impacts on fire-weather in south-east Australia.* CSIRO Australia: Canberra.

Incoll R (1994). Asset protection in a fire-prone environment. In *Fire and biodiversity: the effects and effectiveness of fire management.* Proceedings of the conference, 8–9 October, Melbourne.

IPCC (2007a). *Climate Change 2007: the physical basis – summary for policymakers.* Working Group I contribution to the Intergovernmental Panel on Climate Change 4th Assessment Report.

IPCC (2007b). *Climate Change 2007: impacts, adaptation and vulnerability – summary for policymakers.* Working Group II contribution to the Intergovernmental Panel on Climate Change 4th Assessment Report.

Kerr R (2007). Climate change: pushing the scary side of global warming. *Science* 8 June: 1412–1415.

Kitzberger T, Brown PM, Heyerdahl EK, Swetnam SW & Veblen TT (2007). Contingent Pacific–Atlantic Ocean influence on multicentury wildfire synchrony over western North America. *Proceedings of the National Academy of Sciences of the United States* 104(2): 543–548.

Lindesay A (2003). Fire and climate in Australia. In *Australia burning: fire ecology, policy and management issues* (eds G Cary, D Lindenmayer & S Dovers). CSIRO Publishing: Melbourne.

Lynn K (2003). Wildfire and rural poverty: disastrous connections. *Natural Hazards Observer* November: 10–11.

Main G (2004). Red steers and white death: fearing nature in rural Australia. *Australian Humanities Review* 33. Accessed on http://www.lib.latrobe.edu.au/AHR/archive/Issue-August-2004/main.html on 7 June 2007.

Mendelsohn R, Basist A, Kurukulasuriya P & Dinar A (2007). Climate and rural income. *Climatic Change* 81: 101–118.

McGee TK (2005). Completion of recommended WUI fire mitigation measures within urban households in Edmonton, Canada. *Environmental Hazards* 6: 147–157.

McKenzie FH (1999). *Impact of declining rural infrastructure.* Rural Industries Research and Development Corporation: Canberra.

McLennan J & Birch A (2005). A potential crisis in wildfire emergency response capability? Australia's volunteer firefighters. *Environmental Hazards* 6: 101–107.

Moriondo M, Good P, Durao R, Bindi M, Giannakopoulos C & Corte-Real J (2006). Potential impact of climate change on fire risk in the Mediterranean area. *Climate Research* 31: 85–95.

Mouillot F, Rambal S & Joffre R (2002). Simulating climate change impacts on fire frequency and vegetation dynamics in a Mediterranean-type ecosystem. *Global Change Biology* 8: 423–437.

OECD (2005). *The adaptation landscape.* Organisation for Economic Co-operation and Development information paper JT00193844. Tirpak D(OECD) & Ward M. Global Climate Change Consultancy.

Pearce HG, Mullan AB, Salinger MJ, Opperman TW, Woods D & Moore JR (2005). *Impact of climate change on long-term fire danger.* National Institute of Water and Atmospheric Research, client report AKL 2005–45 for New Zealand Fire Service Commission.

Pitman AJ, Narisma GT & McAneney J (in press). The impact of climate change on the risk of forest and grassland fires in Australia. *Climatic Change.*

Schneider S (2004). Abrupt non-linear climate change, irreversibility and surprise. *Global Environmental Change* 14: 245–258.

Schipper L & Pelling M (2006). Disaster risk, climate change and international development: scope for, and challenges to, integration. *Disasters* 30(1): 19–38.

Smit B & Wandel J (2006). Adaptation, adaptive capacity and vulnerability. *Global Environmental Change* 16: 282–292.

Tarrant M (2006). Risk and emergency management. *Australian Journal of Emergency Management* 21(1): 9–14.

Thomas CD, Cameron A, Green RE, Bakkenes M, Beaumont LJ, Collingham YC, Erasmus BFN, Ferreira de Siqueira M, Grainger A, Hannah L, Hughes L, Huntley B, van Jaarsveld AS, Midgley GF, Miles L, Ortega-Huerta MA, Peterson AT, Phillips OL & Williams SE (2004). Extinction risk from climate change. *Nature* 427 (6970): 145–148.

Victorian Government (2007). *Climate change and infrastructure: planning ahead.* Report of the Victorian Climate Change Adaptation Program. State Government of Victoria: Melbourne.

Index

ABC Local Radio Victoria 115
AFAC Policy see prepare, stay and defend or leave early (AFAC Policy)
Alpine fires (2003) 42
arson 5, 99–106
arsonists
    adult 101–3
    juvenile 103, 105
    punishing and treating 104–5
Ash Wednesday Bushfire Review Committee 61–2
Ash Wednesday fires 78
Australian Capital Territory 21, 66–7
Australian Interagency Incident Management System (AIIMS) 114
Australasian Fire Authorities Council (AFAC) 3, 54, 59, 78, 87
    policy 60–5

building safety 70, 77–85
bushfire community safety 3–7, 47, 59, 129–37
Bushfire Co-operative Research Centre (CRC) v, vii, ix, 3–4, 24, 47, 48, 150, 161
Program C vii, 3, 4, 131, 150
bushfire preparation activities 12, 15, 31
bushfire preparedness 47, 60, 67, 68–9, 85, 117–26
    and children 55–6
    and CFU members 32
    household 117
    implications for public education 119, 121–4
    public education programs 117
    socio-economics 53
    understanding reasons for not preparing 118–23
bushfire risk see community perceptions of bushfire risk

bushfires
    causes 121–2
    and climate change 178, 179
    community safety initiatives 139–50
    community safety programs 129–37
    converting motivation to action 123–4
    damages from 153–4
    differential experience with 50–3
    economic analysis 157
    and economics 153–4
    experience and expectation of outside help 52
    experience and resources 42
    fatalities 5, 48, 59, 60, 61, 62–5, 67, 70, 80, 84–5, 90, 95, 153, 164, 171, 176, 179
    and gender 51–2, 53, 55
    implications for public education 119, 121–4
    management, cost of program 156
    management, costs and benefits 155
    management, economics of 151–9
    perceived importance of 122–3
    public education programs 117
    resilience 176–7
    resilience impacts and communities 178–82
    resource allocation and fire management programs 158–9
    responsibility for action 121–2
    roles between household members 51–2
    risk perception 12, 15, 16–17, 29, 120
    simulations 79
    threat, understanding and assessment 79–81
    transferring responsibility to other community members 122
    trauma 69–70
    vulnerability 176–7
    and weather 175–6

see also economics of bushfire management; fires; prepare, stay and defend or leave early (bushfire strategy)

California ix, 17, 22–3
Canberra fires (2003) 66–7
case studies *see* community case studies
children and fire preparedness 55–6
climate change 6
    adaptation 182
    and bushfires 178, 179
    and climate variability 181–2
    and community bushfire resilience 175–82
    and ecosystems 181
    and fire vulnerability 175–7
    and rural landscapes 180
commitment to stay 68
communities
    and bushfire resilience 178–82
    disaster-resilient 182
    fire experience variability in 53
    local knowledge and rural 38
    and media warnings 113–15
community case studies 13–18, 133–7
    Community Fire Units (CFUs) 133–5
    community survey data collection 14
    Street FireWise (SFW) (Rural Fire Service) 135–7
    Tamborine Mountain 14–16, 18
    Thuringowa 16–17, 18
community consensus and local knowledge 42–3
community engagement and local knowledge 44
Community Fire Units (CFUs) 5, 21, 23–34
    case studies 133–5
    community integration 31
    and community resilience 30
    and community safety 29–30
    involvement from high-risk groups 32
    local knowledge 30
    member focus groups 26
    member questionnaire 26
    members and bushfire preparedness 32
    membership management 33
    methodology 25–6
    operational logistics 31
    perceived risk 29

    prepare, stay and defend or leave early policy 30, 32
    research and policy 33
    social capital 29
    stay and defend or leave early policy 31
    survey findings 26–8
    training and development 31
    who joins and why? 29
community integration 31
community involvement in hazard management 18–19, 22, 67
community perceptions of bushfire risk 11–20
    community involvement in hazard management 18–19, 22, 67
    differential experience with fire 50–3
    evaluating (experts and the public) 18
    need for local-based approach 18
    practical outcomes and recommendations 20
    see also risk
community programs 6, 22
community responsibility 18–19, 22
community safety 3–7, 47, 53, 56, 59
    and CFUs 29–30
    initiatives 139–50
    program logic models 141–2
    program planning and evaluation 148–50
    program theory 129–37
    structured concept mapping 142–8
community survey data collection 14
concept mapping, structured 142–8
contingency planning 68
Country Fire Authority (CFA) 38, 109, 135, 153, 182

damage to assets and outputs, valuing 158
Dandenong Ranges fires (1997) 61
deaths *see* fatalities, bushfire
defend or leave early policy *see* prepare, stay and defend or leave early (AFAC policy)
defendable space 22, 60
Department of Sustainability and Environment (DSE) 109, 110–14, 115, 116, 154
disaster-resilient communities 182
disaster risk reduction *see* risk, reduction
duty of care 90–1

economic theory 154–7
economics and bushfires 153–4
economics of bushfire management 151–9
  economic analysis, limitations for 157
  economic impact assessment 158
  resource allocation and fire management
    programs 158–9
  science for decision making 151–2
  valuing damage to assets and outputs 158
  valuing environmental costs and benefits
    158
  valuing resources affected by bushfires 157
  valuing resources used in fire
    management 157
emergency services 5, 30, 53, 59, 60, 69, 90,
  109, 112, 115, 162, 163, 166
emergency services organisations (ESOs)
  87–8, 93–5
environmental costs and benefits, valuing 158
evacuation powers 88–9
evacuations
  early 93–4
  last-minute 93
  see also prepare, stay and defend or leave
    early (bushfire strategy)
evaluation *see* community safety
Eyre Peninsula fires (2006) 48, 55, 63

fatalities, bushfire 5, 48, 59, 60, 61, 67, 70, 80,
  84–5, 90, 95, 153, 164, 171, 176, 179
  database 62–5
  and late evacuation 84–5
fencing and house vulnerability 83
fire agencies and local knowledge 38
fire management programs
  providing additional resources to 159
  resource allocation and 158–9
fire planning
  local knowledge 5, 22, 23, 30, 31, 35–45,
    102, 137, 146
fire services and media 107–116
  ABC Local Radio Victoria 115
  and community warnings 113–15
fire and land management agencies in
  Victoria 110–13
fires
  differential experience with 50–3
  experience and expectation of outside
    help 52

experience and resources 42
experience variability in the community
  53
and gender 6, 51–2, 53, 55, 167–8
prevention 5, 22
risk perception 12, 15, 16–17, 29, 120
roles between household members 51–2
see also bushfires; 'prepare, stay and
  defend or leave early' bushfire strategy

gender and fires 6, 51–2, 53, 55, 167–8
government departments and local
  knowledge 38

hazard management, community
  involvement in 18–19, 22, 67
Hobart fires (2006) 66
home as refuge 47, 54–5
  see also 'prepare, stay and defend or leave
    early' bushfire strategy
house vulnerability 81–3
  effects of surroundings on house loss
    82–4
  ember accumulation 81
  ember entry (common gaps and entry
    points) 81
  external doors 82
  and fatalities 84–5
  fencing 83
  glazing systems 82
  influence of human behaviour 84–5
  integrated approach 85
  outbuildings 83
  outcomes and recommendations 85
  radiant heat from surroundings 81
  roof, eaves and fascias 82
  surrounding vegetation 83
  timber decking 82
  vulnerable building parts 81–2
  water tanks 83–4
  see also property safety
human behaviour and house vulnerability
  84–5
Hyogo guidelines 21

institutional issues 4, 6, 70

last-minute evacuations 5, 61, 70, 93, 95
leave early, clarification of 69
legal actions 89–90, 91–2

legal issues (defend or leave early policy) 60,
70, 87–96
duty of care 90–1
evacuation powers 88–9
immunities 91–3
torts: negligence 90–1
legislation 87–9, 93
local knowledge 5, 22, 23, 30, 31, 35–45, 102,
137, 146
Alpine fires (2003) 42
applications of 40–2
community consensus 42–3
community engagement 44
and existing community groups 40
and fire agencies 38
and government departments 38
harnessing 38–9
idea of 37–8
key players 40
pitfalls and potential dangers 42–4
resourcing 40
rural communities 38
strengths and weaknesses 41
use of language 43

media 5
ABC Local Radio Victoria 115
and community warnings 113–15
and fire services 107–16
and fire and land management agencies in
Victoria 110–13

New South Wales fires 21–34
North-east Victoria fires (2003) 65

Otway Ranges fires (1983) 62

post-bushfire surveys 78–9
prepare, stay and defend or leave early (AFAC
policy) 5, 22, 24, 25, 30, 31, 32, 47, 54, 55,
59–71, 87–96
building safety 70
bushfire education 67–8
case studies 65–70
commitment to stay/contingency
planning 68
decision making 68
defendable space 60
emergency services 69
historical evidence of the policy 61–2

home as refuge (Wangary, 2005) 47, 54–5
implementation issues 65–70
knowledge of policy 67
leave early, clarification of 69
legal issues 60, 70, 87–96
outreach activities 67–8
policy 33, 59, 60–5, 93–5
preparedness 68–9
promotion of 94–5
trauma 69–70
program logic models 141–2
program logic/program theory 130–3
program planning and evaluation 148–50
program theory and community safety
129–37
case studies 133–7
property safety 70, 77–85
bushfire threat, understanding and
assessment 79–81
methodology 78–9
outcomes and recommendations 85
post-bushfire surveys 78–9
simulations, laboratory and full-scale 79
see also house vulnerability
public education and bushfires
causes of bushfires and responsibility for
action 121–2
converting motivation to action 123–4
factors that motivate people to act 120–1
implications for 119, 121–4
perceived importance 122–3
preparing for bushfires 120
programs 117
transferring responsibility to other
community members 122
understanding reasons for not preparing
118–23

resilience see bushfires, resilience;
communities, and bushfire resilience;
Community Fire Units (CFUs), and
community resilience
resource allocation and fire management
programs 158–9
risk
community involvement in hazard
management 18–19, 22, 67
evaluating (experts and the public) 18
perception 12, 15, 16–17, 29, 120

reduction 22, 85
social construction of 11–12
understanding community perceptions
    18–20
see also community perceptions of
    bushfire risk
rural communities and local knowledge 38
Rural Fire Service 135–7
rural–urban interface ix, 21–34, 41, 64, 70,
    77, 78, 85, 118, 163

self-sufficient communities 3, 4
social construction of risk 11–12
South Australia fires 47–56, 78
Standard Emergency Warning System
    (SEWS) 69
stay or go *see* prepare, stay and defend or
    leave early (AFAC Policy)
stay and defend or leave early *see* prepare,
    stay and defend or leave early (AFAC
    Policy)
Street FireWise (SFW) (Rural Fire Service)
    135–7
Sydney bushfires (2001–02) 4

Tamborine Mountain (community case
    study) 14–16, 18
Thuringowa (community case study) 16–17,
    18
traditional ecological knowledge (TEK)
    37–8
trauma, bushfire 69–70

urban interface ix, 21–34, 41, 64, 70, 77, 78,
    85, 118, 163

vegetation, house vulnerability and
    surrounding 83

Victoria, fire and land management agencies
    110–13
volunteers 23, 33, 38, 69, 70, 91–2, 100, 102,
    110–11, 154, 157, 180
member commitment 164–8
new volunteers 168–71
recruiting and retaining 6, 161–72
training 31
women 167–8
vulnerability 15, 21, *see also* bushfire
    vulnerability; house vulnerability;
    resilience

Wangary fire (January 2005) 47–56
children and fire preparedness 55–6
community safety 47, 53, 56
differential experience with fire 50–3
fire experience and expectation of outside
    help 52
fire experience and resources 42
home as refuge 47, 54–5
household characteristics 49
methodology 48
orienting values and concepts 50
roles between household members 51–2,
    53, 55
water tanks and house vulnerability 83
weather and fires 175–6
Wilson's Promontory fire (2005) 111–13, 116
women
    role of 6, 51–2, 53, 55
    volunteers 167–8
World Conference on Natural Disaster
    Reduction (WCNDR) 21